Creating and Managing a Technology Economy

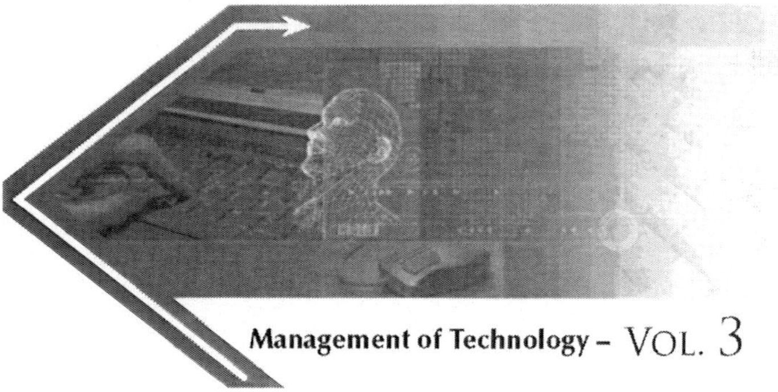

Creating and Managing a Technology Economy

edited by

Fredrick Betz
Nile University, Egypt

Tarek Khalil
COB, IAMOT

Yasser Hosni
President, IAMOT

Hosam Eldeen Mostafa
Nile University, Egypt

World Scientific

NEW JERSEY • LONDON • SINGAPORE • BEIJING • SHANGHAI • HONG KONG • TAIPEI • CHENNAI

Published by

World Scientific Publishing Co. Pte. Ltd.
5 Toh Tuck Link, Singapore 596224
USA office: 27 Warren Street, Suite 401-402, Hackensack, NJ 07601
UK office: 57 Shelton Street, Covent Garden, London WC2H 9HE

British Library Cataloguing-in-Publication Data
A catalogue record for this book is available from the British Library.

CREATING AND MANAGING A TECHNOLOGY ECONOMY
Management of Technology — Vol. 3

ISBN-13 978-981-4313-38-4
ISBN-10 981-4313-38-6

Printed in Singapore.

Contents

Introduction .... **ix**

PART 1: Innovation **1**

Honda's Predisposition Towards Radical and Disruptive Innovations
W. David Holford and Mehran Ebrahimi 3

Innovation Capability Reconfiguration in Business Transition: A Case Study on Taiwanese PC Firm
Ting-Kuei Kuo and Tim Minshall 33

The Emergence of Wireless Networks
Wenshin Chen and David Bennett 57

PART 2: R&D 79

Technology Strategy of R&D Internationalization: An Empirical Study from a Developing Country
Louis Y. Y. Lu and T. M. Chen 81

Russian R&D Organisations: Cases of International Technology Collaboration
Anna Trifilova 109

PART 3. Services **143**

Building a Global Electronic Manufacturing Service Provider: Vendor's Process in Outsourcing
Zoran Perunović and Mads Christoffersen 145

Support Services in Developing Technology b-to-b Relationships
Juha-Pekka Koistinen and Pekka Eskola 165

Designing Rapid Services for Competitive Manufacturing
Deepak A. Sachdev and Darius P. K. Singh 199

PART 4: Industry 221

Productivity Assessment of Implementing Wireless Technologies in Steel Construction Using Simulation Technology
Amine Ghanem 223

Next Generation PLM - An Integrated Approach for the Development and Management of the Product Service Systems in the Telecommunications Industry
Julius Golovatchev and Oliver Budde 245

Ranking Management of Technology Conferences
Harm-Jan Steenhuis and Eerik J. de Bruijn 265

PART 5: Management **283**

An Empirical Study of Information System for Disruption Management
R. Abdi and S. Sharma 285

Technology Management: Best Practises of the South African Automotive Supplier Industry
Marthinus P. Fick and André J. Buys 313

Assessing and Improving Project Management Information System in a Multi National Company
Omar Khalifa Gneiber and Gebril Mohamed Zletni 333

Management of Technology Support Center for Enhancing Competitiveness of Small and Medium Enterprises in Egypt
Yasser Tawfik and Tarek Khalil 357

PART 6: Community **379**

Authorities, Hubs, and Brokers in Communities of Practices
Marianne Hörlesberger and Petra Wagner-Luptacik 381

Does a Favorable Environment have a Positive Effect on University Technology Transfer Activities?: A Case Study on Two Texas State Universities
Michi Fukushima 401

Introduction

The IAMOT conference in Dubai in 2008 was interesting in two ways, the content of the presented papers and the context of country. Of the about three hundred papers presented, we have selected seventeen for publication. These papers all presented interesting theoretical issues along with empirical cases to provide evidence about the theory.

We have classified the papers into related categories. In these, one can see an emphasis on the emerging MOT issues concerning services and information technology. Also one can still see some continuing emphasis on manufacturing technology issues. But what is in light of subsequent and recent events about global economic development and global financial crises, one can identify a subject which MOT has not deeply addressed in this and previous conferences. This is an issue about technology and capital.

And this issue about technology and capital was vividly suggested by the context of the conference in Dubai. On arriving in Dubai, for the conference, a visitor was struck by the fact that this bit of desert of the Dubai Emirate was planning its primarily as a financial and trading center for the Arabian Gulf area. Its sister Emirate Abu Dubai had oil, and Dubai had none. But Dubai had embarked upon a rapid and vast construction project building hundreds of office towers and residential towers. The contrast between construction and urban spread and the desert and ocean was amazing.

Then a year later, as this book was going to press, Dubai was in the international news — in a big way. On 27 November 2009, Jennifer Hughes, David Oakley, Simeon Kerr reported: "Stock markets around the world were convulsed yesterday as investors scramble to understand the implications of Dubai World's restructuring and unexpected debt standstill." (Hughes, et al, 2009).

Dubai World was a holding company owned by the Dubai Emirate government: "Banks were scrambling to quantify potential losses in Dubai after Dubai World, the state's holding company, shocked creditors by asking to halt debt repayments... Three years ago, demand was ... strong for the Nakheel bond at the centre of the Dubai's troubles... Yesterday (26 November) those bonds were trading just above 70 cents on the dollar..." (Hughes and Oakley, 2009).

The seriousness for Dubai's future as a financial center due to this bond payment default was emphasized: "It came in a short statement about the restructuring of Dubai World; one of the emirates biggest and best-known companies ... the decision to ask bondholders of the company ... to extend maturities from December to May 2010 was a bombshell. And the Middle East's most glamorous and creative emirate will pay the price of its decision for a long time to come." (Khalaf, 2009).

At the time of IAMOT's Dubai conference, a year earlier, we had been struck by the enormous amounts of construction going on and also the apparent emptiness of most of the buildings. We wondered (1) where the finance for such major expansion was coming from (since one knew that Dubai did not have oil) and (2) what was the economic development strategy? A year later, we were learning that the money was borrowed and to an amount beyond having sufficient revenue to meet the debt obligations.

In this context of Dubai's plan for economic growth, one can now ask an important question for future research in MOT. This is whether a purely financial strategy for economic development — without a complementary technology strategy — was really a viable strategy for long-time economic development in the global world?

We think this is an important issue which MOT needs to address more deeply. What are the proper interactions between financial strategy and technology strategy for long-term economic development of a nation? MOT has long addressed this issue at a company level but not seriously at a national and global level.

Hughes, Jennifer, David Oakley, and Simeon Kerr. 2009. "Dubai sends markets into turmoil", Financial Times, November 27, p. 1.
Hughes, Jennifer and David Oakely. 2009. "Jitters spread through Middle East", Financial Tims, November 27, p. 2.
Khalaf, Roula. 2009. "Emirate will pay dealy for a long time to come", Financial Times, November 27, p. 2.

PART 1

"Innovation"

- *Honda's predisposition towards radical and disruptive innovations.*
- *Innovation capability reconfiguration in business transition: A case study of Taiwanese PC firm.*
- *Shaping organizational legitimacy: the emergency of wireless networks.*

HONDA'S PREDISPOSITION TOWARDS RADICAL AND DISRUPTIVE INNOVATIONS

W. DAVID HOLFORD

Department of Management and Technology, School of Management, University of Quebec at Montreal (UQAM), P.O Box 8888, Downtown Station, Montreal, Quebec, H3C 3P8, Canada, holford.w_david@uqam.ca

MEHRAN EBRAHIMI

ebrahimi.mehran@uqam.ca

ABSTRACT:

Honda's long term view to R&D does not seem to hamper its 'nimbleness' to provide innovative products that are not only new to its historical business markets, but offer significant differentiating product attributes. Honda's recognized innovative prowess in most industries that it focuses upon leads us to ask the following two research questions: 1) What type of innovation approach (e.g. sustaining, radical or disruptive) differentiates Honda as being an innovative firm; and 2) What type of organizational approach supports the type of innovation that Honda generates?

A critical hermeneutical analysis of key R&D reports and second-hand interviews related to the HondaJet and Ridgeline product development initiatives followed by a review of past organizational research conducted on Honda as well as certain management literature on innovation generated the following results:

Product development programs showed clear patterns of coherency with technologies and design themes developed in preceding product development projects. Certain pertinent analogies were induced by these 'coherent innovation' patterns: namely, 'living' mythologies and societies, 'healthy doubt' or ambivalence in regards to retained knowledge, and creative routines. These analogies are a first indication of dialectics, whether as co-existence or synthesis of contradictions. A further examination of the management literature on Honda clearly indicates Honda's dialectical approach within the organization. Our own cross reference to technical reports and second-hand verbatim in relation to both the HondaJet and the Honda Ridgeline product development programs, as well as to their respective precursor development programs, revealed that the organizational approach can best be described as a dialectical holographic metaphor. There is also a clear relationship between Honda's dialectical holograph, especially in regards to knowledge creation as seen across the HondaJet and Honda Ridgeline development programs, and a general predisposition towards producing radical and disruptive innovations. A conceptual framework is proposed with the aim of providing further insights as to how an organization can achieve the dialectic holographic metaphor; and in turn, a general predisposition towards generating either radical or disruptive innovations.

Conclusions:

1) *Both the HondaJet and the Honda Ridgeline are viewed as being both radical and disruptive innovations. Furthermore, both product development programs showed clear patterns of coherency with technologies and design themes developed in preceding product development projects (hence, 'coherent innovation'). Hence, Honda's resulting innovations are simultaneously characterized by both continuity and novelty.*
2) *Honda's organizational approach in support of this type of innovation can best be described as a dialectical holographic metaphor.*
3) *Honda's approach is also in marked contrast to certain established literature as to how to attain disruptive innovations.*

Keywords: *Radical innovation, disruptive innovation, dialectics, knowledge creation, knowledge sharing*

1.0 Introduction

Historically, the Honda Motor Co. has been somewhat of an enigma to various researchers in management – many contradictory positions seemed apparently justified when looking at various aspects in isolation. Its pursuit of R&D activities appears to pose the same paradox. Honda's long term view to R&D does not seem to hamper its 'nimbleness' to provide innovative products that are not only new to its historical business markets, but offer significant differentiating product attributes. For example, Honda's first experience and resounding success in the North American motorcycle market in the 1970's is described by Christensen (1997) as an example of having developed a disruptive technology across its 50 cc motorbike; while Honda's first pick-up truck (the Ridgeline) unveiled in 2005, won the NAIAS North American Truck of the Year Award in early 2006 with better than expected sales. And finally, its recent announcements to enter the microjet market across the commercialisation of its HondaJet aircraft appear to be consistent with Honda's history of never entering new markets with a "me-too" approach.

This conceptual paper first examines the evolution of the technical aspects and technologies that led to the differentiating attributes of the HondaJet and the Honda Ridgeline pick-up truck by conducting a critical hermeneutic analysis of public documents related to Honda's aerospace and automotive/pick-up truck R&D. Honda's R&D 'strategy' and activities are indeed coherent, but within a 'holographic' dialectical movement. We will also show that this movement provides a predisposition towards generating disruptive innovations as defined by Christensen (1997). We then propose a conceptual framework which identifies the enabling conditions and organisational dynamics that are required towards attaining Honda's 'holographic' dialectical movement.

2.0 The HondaJet

The HondaJet was first unveiled in December of 2003 and is described as being able to seat up to 6 people with a maximum speed of 420 knots and service ceiling of 41,000 ft. The engines, being optimally positioned on the upper surface of the main wing in a unique configuration, reduce drag at high speeds and increase cruising efficiency. This also eliminates the need for structural engine mounts in the fuselage, which according to Honda, creates "30% more cabin space than in conventional aircraft" such as the Cessna Citation 550, the Gulfstream 100 and the Learjet 55 (Honda Worldwide, 2003). The main wing features aluminium skin panels formed from single sheets of aluminium to provide a smoother than conventional surface. This, combined with Honda's proprietary low-drag laminar flow wing (SMH-1) and fuselage nose sections work together to significantly improve aerodynamic performance. The fuselage's compact and lightweight carbon composite and honeycomb sandwich structure also provides added gains in interior space. All these features allow the aircraft to achieve fuel efficiencies up to 40% higher than conventional aircraft. In this regard, the HondaJet, with its radical improvement in existing product attributes of cabin space and fuel efficiency over other conventional aircraft, appears to represent a radical (as opposed to a disruptive) innovation in the spirit of Christensen's (1997) definition (disruptive innovations involve the introduction of new product attributes, while radical innovations involve step-change improvements in existing product attributes).

The evolution of these specific technologies can be traced back over three sequential aircraft research programs over the past 20 years. In 1986, a joint research agreement was first initiated between the Mississippi State University (MSU) and Honda R&D involving a Beech A-36 aircraft that was to be modified with a composite structure (Nakayama and Bennett, 1994). This was followed in 1989, with the MH-02 joint research aircraft project between MSU and Honda R&D, resulting in a prototype 6-passenger jet aircraft in 1992 (Nakayama and Bennett, 1994). It was described as being "the first all-composite small business jet, using lightweight carbon fibre reinforced epoxy resins in all the structural elements" (Honda Worldwide, 1997). As for the HondaJet itself, development began in the late 1990's.

2.1 Coherencies with precursor programs

A certain temporal coherency and continuity between the three aircraft development programs was discerned in terms of technology and design themes to the point that we can present it as being a series of 'coherent innovations'. Firstly, from a technology aspect, the converted Beech A-36 research program involved the learning and gaining of experience of composite material design, fabrication and testing (Nakayama and Bennett, 1994). Past rudimentary knowledge from the automotive sector, along with new trials and experimentation on the Beech A-36, allowed the MSU/Honda team to further learn "the basics of composite material technology and the basics of aircraft design, fabrication, and testing" (Nakayama and Bennett, 1994). This newly attained composites material knowledge, along with further advances in automotive experience in composite materials (resin mould fabrication), was subsequently used within the new context of the MH-02 forward swept-wing aircraft development program (Nakayama and Bennett, 1994; Sullivan et al, 1994). Hence, new composite materials knowledge was created in regards to all the structural elements of the aircraft ranging from main and tail wing cross beams and ribs, to fuselage frame and other outer panels of the aircraft. In a similar manner, composite materials knowledge acquired was subsequently carried forward into the context of the HondaJet program to thereby produce new knowledge with respect to the fuselage (Fujino, 2003; Black, 2006).

Secondly, the above-wing engine mounts were first seen on the MH-02 and were carried forward onto the HondaJet. Furthermore, the MH-02/HondaJet teams use of above wing engine mounts was inspired from decades-old technology found on such planes as the Fokker VFW-614. Yet, this technology, originally chosen for increasing cabin space, could not be used without synthesising it to new ideas, since in its original form presented poor drag characteristics. Across optimal engine nacelle and pylon positioning and design, what was originally a disadvantage eventually became a strong point, whereby drag characteristics were rendered *superior* to conventional under-the-wing designs (Fujino, 1994 and 2003).

Thirdly, an automobile design theme was discerned in both the MH-02 and the HondaJet: Fujino (1994) describes how one of the design requirements of the MH-02 was that crew and passengers could easily get on and off the aircraft without any steps, "just like an automobile"; and the chief test pilot mentioned "the crew seating is very low to the ground and provides a 'sports car' view and feel" (Wilson, 1995). As for the HondaJet, Honda's president Fukui explained "We have created a business jet with high performance, high fuel efficiency, low emissions and a spacious cabin. Sounds like a Honda, right?" (Phelps, 2004a).

The manner in which radical improvements in existing product attributes were achieved helped produce a new product attribute (relative to other competitors): namely, the car-like comfort and handling to the aircraft. Aircraft manoeuvrability coupled with other technologies and design themes already described also contributed towards producing this new and therefore disruptive product attribute. In this sense, the HondaJet can be viewed as being both a radical and a disruptive innovation in the spirit of Christensen's (1997) definition (Figure 1).

Radical
- **40% superior fuel efficiency***
- **30% superior cabin space***

***: Cessna Citation 550; Gulstream 100; Learjet 55**

Disruptive
- **Car-like comfort and handling**

(Low-to-ground design, all-glass cockpit, large cabin space, maneuverability, etc.)

Figure 1. Radical and disruptive product attributes of the HondaJet

3.0 The Honda Ridgeline

Between its first launch (2005) and having won the NAIAS 2006 North American Truck of the Year Award, Honda's first pick-up truck was described by numerous automotive reviews as a vehicle that "doesn't look or act like any other pickup truck we've ever driven"..." is one of the nicest midsize trucks driven in terms of comfort and ease of use"..."sets a new paradigm in terms of body structure for a pick-up", whereby "Honda steps up and redefines the genre with... the most innovative vehicle in a long time" (McGraw, 2005; Vanderwerp, 2005; Truett, 2005).

Innovative highlights of the Ridgeline include: 1) integrated cab and bed unibody construction, resulting in a body-chassis structure with 20X the tensional stiffness and 2.5X the bending stiffness of a competitive body-on-frame truck; 2) first four-wheel, fully independent suspension system to deliver a superior car-like ride and handling, while maintaining the conventional strength of a truck; 3) the industry's first in-bed trunk, made possible across the elimination of the full rear-axle; and 4) a dual action tailgate, hinged at the bottom and side for drop-down and swing-out operations, thus giving easier access to the trunk (Honda Canada, 2005).

The differentiating attributes of the Ridgeline can be summarised as: a mid-size truck offering superior cabin and rear-bed space with a loading capacity matching large-sized pick-ups; coupled with car-like comfort and handling (McGraw, 2005; Vanderwerp, 2005; Truett, 2005). Superior improvements in existing product attributes (e.g. strength and stiffness), along with the incorporation of new car-like performance attributes (comfort, handling, cabin space) relative to existing competitive pick-up truck attributes, allows us to also consider the Ridgeline as both a radical and disruptive innovation (Figure 2) as defined by Christensen (1997).

Radical ➢ **20x torsional, 2.5x bending stiffness vs conventional body-on-frame**

Disruptive ➢ **Car-like comfort and handling**
(independent suspension, 5-speed auto-transmission with 5000 lbs towing capacity, in-bed trunk, car-like cabin space and comfort, etc.)

Figure 2. Radical and disruptive product attributes of the Honda Ridgeline

3.1 Coherencies with precursor programs

A core team of 37 engineers, led by project engineer Gary Flint, did the engineering design work on the truck in a little over four years. The total development time was five years. During this period at least two separate prototypes were built prior to attaining the final design product. One paradoxical comment, comes from Lindsay Brooke, a senior analyst at CSM Worldwide, stating that "For decades, they've [Honda] mastered the art of developing new vehicles without creating all-new ones, using many carryover parts and common assembly processes" (Truett, 2005). Brooke's comments are certainly valid when considering that the starting point of many of the Ridgeline's innovative features such as the unibody construction and the car-like suspension and transmission began with the Honda Pilot and Honda MDX (both SUV's). In this respect we see a continuity with Honda's past experience thus resulting in 'coherent innovation' as discerned in our analysis of the HondaJet. Yet, although the Ridgeline's integrated body-on-frame chassis construction stems from the Honda Pilot SUV, the structure has been significantly changed, with 93 percent being unique to the new truck. The Pilot's five-speed automatic has been extensively reengineered for pickup duty, helping to give the Ridgeline a 5000-pound towing capacity. The front-strut, rear-multilink suspension design is also carried over from the Pilot, but it, too, has been strengthened to handle additional loads (Vanderwerp, 2005). In a similar way, the Ridgeline team tapped into old 'station-wagon' technology for the dual action tailgate. But its original configuration, presented significant load strength disadvantages. Hence new knowledge was introduced so that, according to Flint, "The tailgate is so strong, I can back two trucks up to each other and I can park another Ridgeline on top of them" (Peter, 2005).

4.0 A Few Pertinent Analogies to Honda's Coherent Technologies and Design Themes

The coherent aspects of technologies and design themes which culminated in the HondaJet and Ridgeline remind us of pertinent analogies. Within both development programs, past and existing knowledge was used as a starting point, yet could not be used as is without certain transformations so as to address the new contexts the development programs faced.

Enriquez's (1992) explanation of 'living' mythologies which found in societies that evolve and adapt themselves to changing environments reminds us of Honda's 'coherent innovation'. Such mythologies simultaneously (and continuously) view themselves as being both 'truths' and 'non-truths'. The 'truths' are the aspects of past and existing knowledge which apply to a new situation or context, while the 'non-truths' are the aspects of past and existing knowledge which are no longer pertinent to the new situation or context at hand. It is the mythology's (or society's) self-recognition of these 'non-truth' that opens the way to new ideas and concepts, which in turn, aid it in addressing environmental change. Hence, the mythology (or society) remains 'living'. Conversely, 'dead' mythologies (or societies) are those which consider themselves as being 'absolute truths'. Past and existing knowledge (which are identical, in this case) have no dimensions of 'non-truths', and as a result fail to embrace new ideas or concepts that can aid it in addressing new environmental change or perturbations (Enriquez, 1992).

In a similar vein, Weick (1979 and 2001) speaks of 'living' organisations and "high reliability organisations" which maintain a relative balance between 'crediting' and 'discrediting' its retained knowledge (past and current practices, experiences, logics and beliefs). Organisations that always 'credit' their retained knowledge are those which are non-innovative and fail to adapt to change, while those that that always 'discredit' their retained knowledge fail to learn from past experience, and often end up 're-inventing the wheel'. In order to maintain a proper balance between 'crediting' and 'discrediting', organisations must maintain a "healthy doubt" or "ambivalence" towards its own current beliefs and understandings (Weick, 1979 and 2001).

A third analogy to Honda's 'coherent innovation' is that of "creative routines" defined by Nonaka and Toyama (2002). As opposed to 'frozen' or inertia setting routines, creative routines synthesise both standardisation (or efficiency) and creative change (innovation).

The above-three analogies illustrate a certain ambidexterity, that is, exploitation in co-existence with exploration. It also illustrates 'symptoms' of dialectics of opposing elements (in our case, past/current knowledge versus new knowledge). This can either tend towards the more Western Hegelian type of dialectics involving a synthesis of opposing elements (in our case, creative routines), or the more Eastern type of dialectics involving the co-existence of contradiction (e.g. ambivalence or healthy doubt towards retained knowledge) – Peng and Akutsu, 2001. This tentative theme of dialectics on our part brings us to the following sub-section which attempts to shed further light on Honda's organisational approach.

5.0 The Honda Motor Company as a Dialectical Holographic Metaphor

Most researchers would agree with Cooper (2005) that Honda is an "industry innovator". Yet researchers from different schools of strategic management thinking have used Honda to support a variety of theses. Mair (1999) critically analysed much of this previous literature and argued that Honda had been cited to illustrate "apparently contradictory positions on a series of conceptual dichotomies, namely analytical planning versus learning, market positioning versus resource-based and...core competencies versus core capabilities". He explained that Honda's management thinking is based on the reconciliation of conceptual dichotomies which in turn produced Honda's dichotomy-reconciling strategic capability, as well as being a "specific route to innovation". This argument rejoins Pascale's (1990) description of Honda's management as being able to use paradox and contradiction as an explicit management tool. Nonaka et al (2002 and 2004) go further by providing an explicit explanation as to how contradiction within an organisation is embraced and harnessed in a dialectical fashion (often citing Honda as an example) leading a continuous process of knowledge creation, innovation and overall business strategy. The following examples of Honda's synthesising of opposites across various epistemological and ontological levels of the firm reminds us of Morgan's (1986) "holographic metaphor" in regards to a dialectical system: that is 'the whole can be found within its parts'.

5.1 A knowledge vision which synthesises Honda's absolute and relative value systems

The knowledge vision provides an important criterion for judging the value of knowledge perceived or created, and fosters a context of organisational *commitment* towards creating and circulating knowledge across the organisation (Nonaka and Takeuchi, 2004). According to Nonaka and Toyama (2002), Honda has a knowledge vision synthesising both the absolute and relative value systems of the firm. On the one hand, it emphasises a vision which is not affected by outside conditions to ensure that it can achieve consistency and uniqueness in its knowledge creating activities. On the other hand, Honda cannot ignore the environment with which it interacts. It needs to adjust its activities to the changes in the environment (competitors, etc.). Across the words of Honda president Fukui, we discern a knowledge vision which synthesises both aspects: "Mobility is a basic desire, right and joy for all people. Honda continues to pursue mobility from all dimensions. Mobility in the third dimension – the sky – was a dream we have had since shortly after our company was founded. The words of our founder, 'Do not imitate others' are burned in the minds of everyone at Honda. We have no interest in following." (Phelps, 2004a). All generated knowledge will, on the one hand, be judged against the absolute criteria of mobility and uniqueness, and on the other hand, against the relative criteria of pursuing 'the third dimension – the sky'. The HondaJet with 30% more cabin space and 40% higher fuel efficiency than conventional aircraft clearly meets both criteria.

5.2 Synthesis of realistic and idealistic principles

According to Nonaka and Toyama (2002), Honda uses "two contradicting principles" in daily practice. One is the emphasis on the importance of reality or 'go to the actual place, know the actual situation, be realistic'. For example, Gary Flint, in the development of the Ridgeline, refers to the Japanese term *sangen-shugi*, which means "go to the spot", which he used towards conducting his own market research on customer wants and needs (which was conducted over and above the traditional listening to customer focus groups). The other principle is emphasis on the importance of the ideal by respecting 'sound theory'. In this regard, the Ridgeline development team used the theory behind unibody construction to gain additional resistance to bending and twisting. Yet, as Truett's (2005) explains, "Honda engineers can't just show their bosses diagrams and drawings. They have to design parts that solve problems".

Honda also synthesise practice and theory across three levels of questions (Nonaka *et al*, 2002). One level will look at specifications such as 'what is the aircraft cabin space?' etc. A higher level will look at conceptual and design issues such as 'what is the design of the aircraft cabin?' The highest level will look at basic essential questions, such as 'what is this aircraft for?' When a problem cannot be solved, Honda engineers go up a level in terms of questioning: so if the cabin design cannot be agreed upon, they go back up to 'what is this aircraft for?'.

5.3 Synthesis of consensus/respect and open debate

Kiernan (1996) describes how at Honda the constant questioning of ideas, decisions and management is encouraged, even demanded of each employee. Kiernan adds, "Design and development teams are deliberately staffed with engineers from peripheral disciplines who are unfamiliar with the core technology under development. This is to ensure that problems will be approached from different and innovative perspectives, and that conventional wisdom will be challenged and tested." Along these lines, Kiernan also refers to

Pascale (1990) in regards to Honda's art of 'constructive contention' or *Waigaya* – a contention management protocol where people listen to one another, and can disagree without being disagreeable. This requires management being skilful at facilitating "so as to surface thoughts and feelings present and move toward constructive action" (Pascale, 1996). Hence, *Waigaya* can be viewed as a synthesis of open debate and consensus/respect. Such a synthesis is discerned within the Ridgeline's project development team who "debate, discuss, examine and analyse the engineering and marketing data and from that, they implement a comprehensive product plan. Then they set to work. Once a decision is made, that's it - unless, of course, a team member comes up with new and compelling data. Then the action plan can be revisited" (Cato, 2005).

5.4 Synthesis of different ideas: past vs present, internal vs external, existing vs new

Nonaka and Toyama (2002 and 2004) explain that idea creation can be achieved across the synthesis or integration of opposing aspects through a dynamic process of dialogue and practice. This is precisely what is seen across the team dynamics throughout the development of the Ridgeline pick-up truck, and can also be deduced across some of the R&D documents describing Honda's aviation development activities. The opposing aspects of interest in this case consist of different ideas being discussed and debated, prior to consensus. For example, some of these ideas can be categorised as existing 'in-house' vs existing 'external' ideas, across the Ridgeline team's examination of in-house experience in relation to both the Honda Pilot and MDX in regards to unibody construction, suspension and transmission vs. the teams examination of competitive 'best practices', otherwise known as *zembara* (which Flint explains as "where we took competitive vehicles apart, down to the last bolt, and took note of their construction -- who had the best technique" – Peter, 2005). This exercise culminated into two prototypes being built to test vehicle dynamics and validate all of the modelling and development work, or as Flint said, "to make sure we were going down the right path...The first one...had a [Ford] SportTrac bed in it" (Peter, 2005). The second iteration used a Honda Acura MDX cabin with exclusive side panels.

Both the Ridgeline and MH-02/Hondajet teams also tapped into past knowledge as a source of additional ideas. For example, the MH-02/HondaJet teams use of above wing engine mounts was inspired from decades-old technology found on such planes as the Fokker VFW-614. But this technology, originally chosen for increasing cabin space, could not be used without synthesising it to new ideas, since in its original form presented poor drag characteristics. Eventually, via optimal engine nacelle and pylon positioning and design, what was originally a disadvantage became a strong point, whereby drag characteristics *superior* to conventional under-the-wing designs (Fujino, 1994 and 2003). In a similar way, the Ridgeline team tapped into, and subsequently improved, old 'station-wagon' technology for the "Dual Action" tailgate.

6.0 A Predisposition towards Creating Disruptive Innovation

We already saw across Honda's synthesis of 'contradictions', that a large number of possible dyadic pairs may exist depending on the context at hand (e.g. tacit vs explicit knowledge, external vs internal knowledge, theory vs practice, cooperation vs competition, etc.). Furthermore, each series of differing viewpoints offers a multitude of dyadic arrangements of concepts to be synthesised across dialectical interactions and sense-making between individuals or groups of individuals via dialogue and practice. Hence, complementary oppositions or contradictions are not entrapped in an "either/or" dichotomy, and thereby avoid being limited to only one half of the dynamics within the vast spectrum of dyadic 'pairs'.

It can also be argued that dialectical synthesis also has natural affinities towards achieving disruptive innovations as defined by Christensen (1997). Disruptive innovation involves products that offer new product attributes (as opposed to radical and sustaining innovation which involve products with radical and moderate improvements in *existing* product performance attributes respectively). In focusing on the visible product innovation aspect, we could simplistically say that the past vs present/old vs new/internal vs external series of knowledge related dyads as described in section 5.5 are the ones most directly related towards having produced the disruptive innovations as represented across the Ridgeline pick-up truck and the HondaJet.

Christensen's (1997) illustrates examples of disruptive innovations deriving from "off-the-shelf" components put together in a product architecture that is different and often simpler than prior approaches. Thus, disruptive innovations often involve existing off-the-shelf knowledge and technologies being synthesised with new knowledge (e.g. new architectures, assemblies or interfaces) and/or internal knowledge being synthesised with external knowledge. In many ways, the same can be said for the Honda Ridgeline and the HondaJet, which represented existing knowledge (which in itself involved the synthesis of internal and external knowledge) synthesised with new knowledge (e.g. new ways of putting this existing knowledge together, new unibody construction, optimisation of engine position and nacelles, etc.); hence, resulting in a pick-up truck or aircraft that has all the existing performance attributes of conventional products combined with the added performance attribute of car-like handling, space and comfort.

The dialectical synthesis approach allows for *partial* ruptures with past and existing knowledge, while also allowing for building upon this same existing knowledge; essentially the same pattern discerned within Honda's 'coherent technologies' in sections 2.1 and 3.1. For example, although the Ridgeline's integrated body-on-frame chassis construction stems from the Honda Pilot SUV, the structure had been significantly changed, with 93 percent being unique to the new truck. The Pilot's five-speed automatic transmission had also been extensively reengineered for pickup duty, helping to give the Ridgeline a 5000-pound towing capacity. The front-strut, rear-multilink suspension design was also carried over from the Pilot, but it, too, was strengthened to handle additional loads (Vanderwerp, 2005).

Consistent with Christensen's argument that disruptive innovation need not be costly relative to radical or sustaining innovation, the Ridgeline's project team leader Gary Flint, who worked 15 years at GM prior to joining Honda 12 years ago, explained that "The project's total cost came in well under $250 million. Compared to General Motors, (the Ridgeline's cost) is peanuts," (Truett, 2005). Although we could not obtain direct product development cost comparisons between the Ridgeline and its competitors for 2005 (which is the year the Ridgeline was unveiled), generic 2002 data showed GM to be spending over $400 million per product development project (Welch, 2002). Yet, GM designs were at that time viewed as being relatively dull and "stodgy" (Welch, 2002).

6.1 A Major departure from Christensen's approach

A paradox in this paper is that while we have argued the case that Honda's dialectical, and therefore ambidextrous, approach generates a natural affinity towards producing both radical and disruptive innovations as defined by Christensen (1997), a major epistemological departure from Christensen is worth noting. One of Christensen's (1997) theses is that efficiency and creativity cannot intimately co-exist or interpenetrate one another. Christensen correctly highlights examples of historically successful (and initially innovative) corporations having progressively lost their creative edge across the imposition of large inertia-setting organisational routines. On the one hand, Christensen (1997: 200-202) argues that such an outcome is understandable and even justifiable, in that these routines are meant to be 'frozen' or unchanging in order to maximise efficiency. Christensen's solution for addressing this dilemma is to have the existing organisational entity re-invigorate its creativity by either acquiring or creating a new organisational 'arm'. In this manner, the new 'creative arm' will not be hampered by the existing organisational system. Yet, at a certain stage in growth this new organisational 'arm' needs to establish certain internal efficiencies in order to truly achieve significant levels of productivity and profitabilities. What happens when this 'arm' eventually starts to routinise its own initially innovative successes? We are then obliged to re-start the whole process of acquiring or creating another new 'creative' arm. Christensen's approach implies that any given locus within the overall organisation is either creative or efficient, but not both. This deductive logic not only resembles the deductive 'pendulum swinging' criticised in the previous sub-section, but on meso and more micro organisational level creates barriers to knowledge sharing. One of the strengths of Honda, according to Nonoka and Toyama (2002) is in its ability to synthesise both systemised explicit knowledge with the often more tacit conceptual knowledge (e.g. "creative routines"). This implies both a co-existence as well as a synthesis between efficiency and creativity. Furthermore, this dialectic and ambidextrous approach, as seen in sections 5.1 through 5.6 is seen across all ontological levels of the organisation, thereby producing a holographic metaphor. This is in stark contrast to Christensen's (1997) segregation between creativity and efficiency.

As argued above, Christensen's (1997) proposed methods of achieving disruptive innovation does not appear to concur with Honda's approach. Yet, certain authors appear to have mistaken Christensen's proposed methods with his definitions of what sustaining, radical and disruptive innovation consist of. For example, Nonaka, Kohlbacher and Holden's (2006: 40) adherence to the dialectical synthesis of knowledge (as also proposed in this paper) make the assumption that only sustaining innovation can be produced under such an environment, in that coherent continuity via dialectics is only synonymous with sustaining innovation. Our own examination of Christensen's (1997) definition, as well as the nature of the innovation produced by Honda suggest otherwise. In fact, and as repeatedly argued, we feel that Honda's dialectical and holographic approach, on the contrary, has clear affinities and pre-dispositions towards producing both disruptive and radical innovations.

7.0 Possible Insights as to How to Achieve a Dialectical Holographic Organisation

According to Nonaka and Takeuchi (1995), the synthesising of opposites within Honda is conducted through SECI (socialisation, externalisation, combination, internalisation). Although we do not disagree with Nonaka and Takeuchi's (1995) approach, we believe that a complementary aspect that can further enrich our understanding as to the nature of the dynamics involved in a dialectical organisation resides in examining both group dynamics and organisational sense-making. The group, according to Enriquez (1992: 97 and 101), "constitutes the privileged location for the comprehension of collective phenomena", whereby the "the group only establishes itself around an action to accomplish, a project to work on or a task to complete". When speaking of human interpretation, it is also important to consider the sense-making process as defined by Weick (1995: 14): "the concept of sense-making is valuable because it highlights the invention that precedes interpretation".

An ambivalent attitude (or "healthy doubt") allows for a reasonable equilibrium between trust and disbelief towards past experience and therefore avoids organisations from either becoming ultra-stable or ultra-flexible (Weick, 1979 and 2001). Weick's call for a balance between the 'crediting' and 'discrediting' of retained experiences, beliefs and logics, also rejoins Enriquez's (1992) description of work groups that actively thrive in ambiguity-embracing dynamics (or where contradictions both co-exist and interact with one another): that is, work groups that promote both individual emancipation of its members in conjunction with a visible sense of group identity. Here, both individual emancipation and group identity interact dialectically so as to reinforce one another. Such groups, referred to as *differentiated* groups, is characterised by members who respect and tolerate each other's differences, alongside open dialogue/debate of different viewpoints (hence, another dynamic ambiguity). Enabling conditions for such group dynamics include empathy, autonomy, as well as mutual trust and respect between its members (Enriquez, 1992). Furthermore, preliminary indications from our own on-going qualitative research (interviews and observations within an embedded case study) within an important North American aircraft engine manufacturing company suggests that differentiated groups tend to be open to other group's ideas, all the while not losing their own group identities and ideas. This is also consistent with Stevenson's (2002) concept of local communities interacting with one another within an increasingly global network. Such a community, according to Stevenson (2002: 743), "recognises that, to be creative and viable...it must be inclusive of differences and open to change...Ideas flow in and out. Such a community recognises it has much to learn from others. Complementarities are celebrated". This is on condition that the enabling conditions listed above that are present within the workgroup are also present at the group interfaces (e.g. formal or informal meetings and discussions). In such situations, integration of other group ideas occurs simultaneously with the creation of differentiated ideas within the group, thus keeping a healthy balance between integration and differentiation.

7.1 Required enabling conditions

Nonaka and Takeuchi (1995 and 2004) speak of various conditions that enable organisational knowledge creation and circulation to occur. One of these enabling conditions is, as discussed earlier, instilling a *knowledge vision.* But in order for this knowledge to interact in a manner that involves the continuous re-visiting and re-adjustments of ideas and concepts in a manner that integrates and synthesises the continual input of differing viewpoints and 'contradictions' (or dyads), individuals must first feel secure to interact with one another in an environment of openness, tolerance, and respect. Hence, it becomes primordial that there be an environment in which management nurtures mutual trust, empathy, and care, whereby individuals are more likely to share their personal contexts and knowledge leading to what Nonaka et al (1998) refer to as '*ba*'. But it also means that there is more likelihood for tolerance and openness to differing viewpoints, whereby discussions as well as open debating of ideas can occur within a context of respect and mutual trust. This openness and tolerance on the part of individuals not only allows synthesis of various knowledge dyads to occur (thereby increasing the chances of generating disruptive innovations), but also allows different ontological (individual to group to sub-organisation) and epistemological (tacit and explicit knowledge) dyads to synthesise together thereby leading towards the dialectical holographic firm.

Openness to differing viewpoints coupled with the somewhat more classical enabler of requisite variety (Ashby, 1956) of personal experiences and backgrounds espoused by evolutionary economic and complexity theorists alike (postulating that an organisation's internal diversity must match the variety and complexity of the environment in order to deal with its challenges) increases the likelihood that peripheral ideas or 'weak signals' as described by Day and Schoemaker (2005) are integrated within the knowledge creation process. This again, increases the likelihood of generating disruptive innovations. The fostering of debate and eventual synthesis of differing ideas at Honda is one source of requisite variety. But having people with different backgrounds and functions around a table amplifies this requisite variety. Kiernan's (1996) describes this of Honda: "Design and development teams are deliberately staffed with engineers from peripheral disciplines who are unfamiliar with the core technology under development. This is designed to ensure that problems will be approached from different and innovative perspectives...". Some of the technology leveraging of Honda's existing automotive technologies into the MH-02 and HondaJet aircraft development programs may in part be explained by the presence of engineers from peripheral disciplines involved in both programs.

Added to this is redundancy of information or knowledge whereby there is an existence of information or knowledge that goes beyond the immediate operational requirements of organisational members. This involves the intentional overlapping of information and business activities and management responsibilities. By sharing a concept with other individuals who may not need the concept immediately, more sharing of tacit knowledge is promoted because an individual can sense what others are trying to articulate. In this sense, redundancy of information speeds up the knowledge-creation process. An example would be in concept development whereby individuals are trying to articulate images rooted in tacit knowledge: here redundant information enables individuals to invade each other's functional boundaries and offer advice or provide new information from different perspectives (Nonaka and Takeuchi, 1995). This is reflected in Honda's brainstorming camps, where these "meetings are not limited to project team members but are open to any employees who are interested in the development project underway" (1995: 63). This too, can increase the likelihood of generating disruptive innovations.

7.2 A proposed conceptual framework

A proposed conceptual framework based on what has been presented so far is shown in Figure 3. Differentiated groups are formed and nurtured across the establishment of a knowledge vision that is pertinent to each particular group (this provides a collective vision, will and identity that will aid in choosing and implementing the most promising ideas), care, empathy and mutual trust (to allow members to openly share and debate differing viewpoints without feeling threatened or fear of being ostracised), requisite variety (to enrich differing points of views, ideas and concepts via differing backgrounds, skills and experience as well as by voluntary job rotation), and finally, redundancy of knowledge (to enhance sharing of ideas and viewpoints). Each group's knowledge vision integrates upper management's knowledge vision, while also addressing the realities of the group's specific context. Such teams at Honda are led by middle-management, providing the link between upper management's vision and front-line reality (Nonaka and Takeuchi, 1995). Furthermore, upper management at Honda also act as a team (Kiernan, 1996) in a manner that is very similar to the team-centric approach described by Smith and Tushman (2005). Here, interactions occur, whereby debate is common, yet across frequent interactions make it possible for resolutions, as well as knowledge complexification. Yet, as suggested in section 7.0, the locus of organisational integration is not just limited at the senior level – it occurs across the organisation across differentiated work groups, whereby middle-management play the lead role.

We can argue that a knowledge vision and redundancy of knowledge tends to support the knowledge integration aspect of the organisation, while mutual trust, care, empathy and requisite variety will tend to support the knowledge differentiation aspect of the organisation. In turn, the nurtured differentiated groups, supported by these enabling conditions, will actively generate various types of dialectical processes which originate from their active embracement of dynamic ambiguities such as individual emancipation vs group identity, tolerance and respect vs open debate, and 'healthy doubt' in regards to the groups retained knowledge. Furthermore, by ensuring these same enabling conditions at group interfaces between interacting differentiated groups, the process of dialectics (and thus knowledge creation and sharing) is amplified so as to produce a holographic metaphor. Hence, this holographic metaphor is maintained across a network of interacting teams which have distinctive identities (Hackman, 2002), yet are also open to other groups' input (Stevenson, 2002). In turn, this organisational 'structure' produces affinities towards generating radical and disruptive innovation. Finally, risk is minimised since the 'best of both worlds' is continuously sought, that is efficiencies by integrating lessons-learned across the organisation based on retained knowledge, all the while seeking on-going new knowledge creation so as to continuously adapt to the changing environment. Our definition of risk is, on the one hand, relatively straight-forwards whereby consequences (as an undesirable event) are actual events which originate from inherent risk (Black, 2003); while, on the other hand, we view various risk categories (cost, quality, social, health and safety, etc.) as being inter-related (Perrow, 1984; Beck, 2002). This is based on Latour's (1993) epistemology which sees categories being continually hybridised and inter-related, whereby our traditional Cartesian notion of categories is but a "purified" (and therefore simplified) approximation of reality.

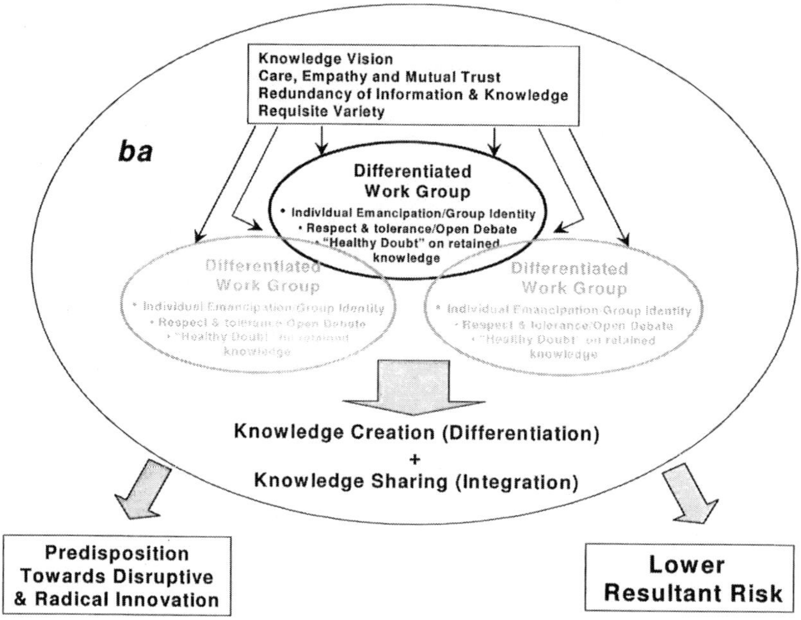

Figure 3. Group dynamics and enabling conditions supporting a dialectic holographic metaphor

How effectively we can reduce risk depends on how well we can complexify our knowledge (Weick, 2001) and thereby minimise the presence of tightly coupled systems (Perrow, 1984). Knowledge complexification is achieved by integrating as many aspects of knowledge as possible (i.e. tacit vs explicit, old vs new, rational vs intuitive/emotional, etc.) – Beck, 2002. The organisation as a dialectical holographic metaphor (as opposed to the linear-causal characteristics of tightly coupled organisations or systems – Perrow, 1984), is an effective way to achieve such knowledge complexification, and thereby reduce risk – all the while increasing our predisposition towards radical and disruptive innovation. This overall outcome, which seems to go against Cartesian 'intuition', tends to be supported by our second-hand analysis of Honda, along with our comparison to the more efficiency-driven firms such as GM and Ford.

Finally, it is perhaps prudent to discuss the issue of transferability in relation to different cultural contexts. Throughout our analysis of Honda, one could argue that Japanese practices are not transferable to Western contexts due to cultural differences. For example, Nonaka and Takeuchi (1995) explain that there is a greater emphasis placed on socialisation within the Japanese workplace when compared to North America. This argument holds a certain degree of validity, but as Kélada (1996: 178) points out "although we must recognise the great difference between American and Japanese cultures, we cannot dismiss all Japanese management practices as being culture oriented and continue to think that anything that does not originate in the United States cannot be applied". For example, Kennly and Florida (1995) argued that Japanese management styles had been successfully transferred over in the automotive industry, involving Toyota (NUMMI) and Honda plants in the United States, and "that there is ample evidence that where Japanese firms have undertaken a concentrated and sustained effort they have been successful in transferring their labour-management system to the USA". Our own hermeneutic study included verbatim and data originating from Honda within a North American context, and seemed to show a reasonable degree of correlation with Japanese literature on knowledge management (Nonaka et al, 2002, 2003 and 2004). Yet, we acknowledge that the issue of culture is complex and highlights the need to conduct empirical studies within Japanese companies such as Honda that are situated within various contexts around the world.

8. Summary and Conclusions

In the first section of this paper, we briefly examined the evolution of certain technologies and design themes that culminated into the HondaJet and the Honda Ridgeline respectively. Both products are viewed as being both radical and disruptive innovations according to Christensen's (1997) definition. Furthermore, both product development programs showed clear patterns of coherency with technologies and design themes developed in preceding product development projects. Certain pertinent analogies were induced by these 'coherent innovation' patterns: namely, 'living' mythologies and societies, 'healthy doubt' or ambivalence in regards to retained knowledge, and creative routines. These analogies are a first indication of dialectics, whether as co-existence or synthesis of contradictions. A further examination of the management literature on Honda clearly indicates Honda's dialectical approach within the organisation. Our own cross reference to technical reports and second-hand verbatim in relation to both the HondaJet and the Honda Ridgeline product development programs, as well as to their respective precursor development programs, revealed that the organisational approach can best be described as a dialectical holographic metaphor. There is also a clear relationship between Honda's dialectical holograph, especially in regards to knowledge creation as seen across the HondaJet and Honda Ridgeline development programs, and a general predisposition towards producing radical and disruptive innovations according to Christensen's (1997) definition. Honda's organisational approach also offers lower overall risks when compared to the more traditional 'deductive pendulum swinging' approach. Yet, Honda's approach is also in marked contrast to Christensen's (1997) own epistemology and proposed process as to how to attain disruptive innovations.

Borrowing from both Enriquez (1992) and Weick (1979 and 1995) in regards to group dynamics and organisational sense-making, and certain associated key enabling conditions, we propose a conceptual framework that we hope can provide further insights as to how an organisation can achieve the dialectic holographic metaphor; and in turn, a general predisposition towards generating either radical or disruptive innovations. Managing an organisation within a dialectical firm, with its very nature of constantly requiring an openness to alternative and differing views, aspects and concepts, requires special emphasis on mutual trust, care and empathy so that individuals and groups feel comfortable and secure in the sharing and exchanging of views and ideas. Furthermore, requisite variety of backgrounds and experience, as well as redundancy of knowledge further enhances this sharing and exchange so as to produce a richer and more complex knowledge able to cope with the complexities and turbulence of the environment. It is this synthesis of various 'contradictions' or viewpoints that gives the firm and its resulting disruptive innovations the simultaneous characteristic of continuity and novelty.

A major weakness of this paper lies in its purely conceptual nature, and that it relies solely on second hand verbatim and technical data. Qualitative fieldwork is currently on-going at a major North American aircraft engine manufacturer towards verifying the validity and veracity of our hermeneutic 'findings'. Future work is also being envisaged towards conducting field work within the Honda Motor Co.

References

Ashby, W.R. (1956), *An Introduction to Cybernetics*, London: Chapman & Hall.

Beck, U. (1992), *Risk Society: Towards a New Modernity*, Newbury Park, CA: Sage Publications.

Black, R. (2003), *Critical Testing Processes,* Boston: Addison-Wesley.

Black, S. (2006), "Unusual one-off a production 'maybe'", Composites World, http://www.compositesworld.com/hpc/issues/2006/January/1157/3

Cato, J. (2005), "Look out Detroit: Honda is now serious about trucks", Canadian Driver, March 29, www.canadiandriver.com/articles/jc/ridgeline1.htm

Christensen, C.M. (1997), *The Innovator's Dilemma,* Boston: Harvard Business School Press.

Cooper, R.G. (2005), *Product Leadership. Pathways to Profitable Innovation,* 2nd Edition, New York: Basic Books.

Day, G. and Schoemaker, P. (2005), Scanning the Periphery, *Harvard BusinessReview*, pp. 135-148.

Dubois, T. (2006), New Engine Technology, Aviation International News, January, www.ainonline.com/issues/01_06/01_06_newenginetech_1.html

Enriquez, E. (1992), *L'Organisation en analyse*, Paris: Presses Universitaires de France.

Fujino, M. (1994), Aerodynamic and aeroelastic design of the experimental aircraft Honda MH02, in 1994 AIAA/FAA Joint Symposium on General Aviation Systems, pp. 435-459.

Fujino, M. (2003), Design and development of the HondaJet, in AIAA Conference 2003-2530, pp. 161-177.

Hackman, J.R. (2002), *Leading Teams: Setting the Stage for Great Performances*, Boston: Harvard Business School Press.

Honda Canada, (2005), Honda Ridgeline Pickup Makes Canadian Debut in Montreal, January 13, www.honda.ca/HondaCorpEng/pressroom/PressRelease/HondaAutomotive/2005-01-13_Ridgeline_Debut.htm

Honda Worldwide, (1997), "Honda 6-passenger experimental jet plane MH02", September, http://world.honda.com/news/1997/c970919e.html

Honda Worldwide, (2005), "Honda and General Electric form strategic alliance in the business jet engine market (update)", January 5, http://world.honda.ie/contentv3/index.cfm?fuseaction=page&pageID=5848

Kennly, M. and Florida, R. (1995), The transfer of Japanese management styles in two US transplant industries: autos and electronics, *Journal of Management Studies*, 32(6), 789-802.

Kiernan, M. (1996), Get innovative or get dead, *Business Quarterly*, 61(1), 51-58.

Latour, B. (1993), *We Have Never Been Modern*, Cambridge, MA: Harvard University Press.

Mair, A. (1999), Learning from Honda, *Journal of Management Studies*, 36(1), 25-44.

McGraw, J. (2005), "Honda builds its first pick-up: and it's a winner", Automotive, www.automotive.com/2005/honda/ridgeline/reviews/index.html

Morgan, G. (1986), *Images of Organisation*, Newbury Park, CA: Sage Publication.

Nakayama, H. and Bennett, G. (1994), Introduction of the MH-02 experimental aircraft: A joint research program of MSU and Honda, in 3rd joint AIAA/FAA Symposium on General Aviation Systems, pp. 103-120.

Nonaka, I. Kohlbacher, F. and Holden, N. (2006), Aging and innovation: Recreating and refining high-quality tacit knowledge through phronetic leadership, in 2006 Annual Meeting of the Academy of Management (AOM), Atlanta, Critical Management Studies Workshop, pp. 1-56.

Nonaka, I. and Konno, N. (1998), The Concept of "Ba": Building a Foundation for Knowledge Creation, *California Management Review*, 40(3), 40-54.

Nonaka and Toyama, (2002), A firm as a dialectical being: towards a dynamic theory of the firm, *Industrial and Corporate Change*, 11(5), 995-1009.

Nonaka, I. and Takeuchi, H. (1995), *The Knowledge Creating Company*, New York: Oxford University Press.

Nonaka, I. and Takeuchi, H. (2004), *Hitotsubashi on Knowledge Management*, Singapore: John Wiley and Sons (Asia).

Pascale, R.T. (1990), *Managing on the Edge*, New York: Simon and Schuster.

Pascale, R.T. (1996), Reflections on Honda, *California Management Review*, 38(4), 112-117.

Peng, K, and Akutsu, S. (2001), A mentality theory of knowledge creation and transfer: Why some smart people resist new ideas and some don't, in *Managing Industrial Knowledge Creation, Transfer and Utilization,* Nonaka, I. and Teece, D.J. (Eds), pp. 105-124, Thousand Oaks, CA: Sage Publications.

Perrow, C. (1984), *Normal Accidents*, New York: Basic Books.

Peter, J. (2005), "Honda Goes Trucking", Automotive News, April, www.ai online.com/adv/Previous/show_issue.php?id=731

Phelps, M. (2004a), "Honda provides more detail about its HondaJet program", Aviation International News, October, www.ainonline.com/issues/10_04/10_04_hodap96.html

Phelps, M. (2004b), "HF118 turbofan is now a Honda/GE joint venture", Aviation International News, March, www.ainonline.com/issues/03_04/03_04_HF118p1.html

Smith, W.K. and Tushman, M.L. (2005), Managing strategic contradictions: A top management model for managing innovation streams, *Organisation Science*, 16(5), 522-536.

Stevenson, T. (2002), Communities of tomorrow, *Futures*, 34(8), 735-744.

Sullivan, R.W., Sato, K. and Bennett, G. (1994), A general review of the fabrication and quality control activities for the Honda MH-02, in 1994 AIAA/FAA Joint Symposium on General Aviation Systems, pp. 489-507.

Truett, R. (2005), "Crafting a pick-up: Ex-GM engineer designed Ridgeline the Honda way", Automotive News, June, www.autoweek.com/news.cms?newsId=102565

Vanderwerp, D. (2005), "For its next swipe at market share. Honda reinvents the pick-up truck", Car and Driver Magazine, March, www.caranddriver.com/article.asp?section_id19&article_id-9144

Weick, K.E. (1979), *The Social Psychology of Organizing*, 2nd Edition, Reading, MA: Addison-Wesley Publishing Co.

Weick, K.E. (1995), *Sensemaking in Organisation*, Thousand Oaks, CA: Sage Publications.

Weick, K.E. (2001), *Managing the Unexpected. Assuring High Performance in an Age of Complexity*, San Francisco: Jossey-Bass.

Welch, D. (2002), "Can Stodgy GM Turn Stylish", BusinessWeek online, November 11, http://www.businessweek.com/magazine/content/02_45/b3807048.htm

Wilson, G.W. (1995), Flight test of a civilian aircraft, in Report to the Aerospace Profession, 39th SETP Symposium, Mississippi State University, pp. 95-113.

INNOVATION CAPABILITY RECONFIGURATION IN BUSINESS TRANSITION: A CASE STUDY ON TAIWANESE PC FIRM

Ting-Kuei Kuo
Graduate Institute of Technology Management
National Taiwan University of Science & Technology
Taipei, Taiwan

Tim Minshall
Centre for Technology Management, Institute for Manufacturing,
University of Cambridge

Abstract

This paper explores the some of the challenges facing computer manufacturing firms in developing the innovation capabilities to allow them to move from Contract Manufacturing (CM) to Original Brand Manufacturing (OBM). As firm transitions from CM to OBM, some of the innovation capabilities may no longer be available or useful while new capabilities may be required to generate the right ideas and respond to market demands. Managing the reconfiguration of innovation capabilities appears to be an important factor in the successful transition from CM to OBM. An initial framework, drawn from concepts described in the literature on the resource-based view (RBV) of the firm, is used to structure a case study of Acer, a Taiwanese computer firm. Acer is used as an example to illustrate the type of challenges faced in moving between CM and OBM, and potential means of overcoming them.

Introduction

There has been an increasing amount of research on the progress and performance of firms in East and South East Asia [1-3]. The unique character of innovation trajectories in East Asia has been highlighted and compared with Utterback and Abernathy's model of industry evolution, as presented in figure 1 [2]. For East Asian firms, the sequence of innovation events is typically reversed, with an evolution from what Utterback and Abernathy defined as mature to the early stages of the innovation process. The Contract Manufacturing (CM) system plays an important role with many East Asian economies, aiding technology transfer and helping firms to attain economies of scale and accumulate learning from process innovation to product innovation [3]. The Taiwanese personal computer industry illustrates this growth pattern particularly clearly [4].

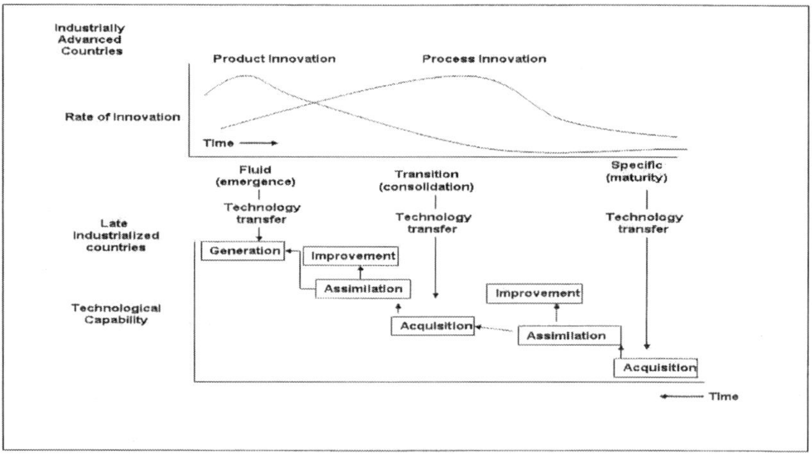

Figure 1. A simple life cycle model of innovation in developing countries [2]

Prior Literature

This literature review covers several relevant areas. Firstly, it discusses what the sources of innovation are and how they might influence a firm. Next, the innovation capability is discussed to elaborate what capabilities are required when moving toward Original Brand Manufacturing (OBM). In section the third section, the mechanisms used by a firm to acquire new capability are reviewed. Both internal and external mechanisms will be discussed. Considering a firm as a combination of resources and capabilities, the resource-based view (RBV) and dynamic capabilities are discussed in the last section. RBV provides a useful starting point to analyze a firm in business transition. Finally a framework synthesizing the relevant concepts is proposed.

Sources of innovation

Innovation is a widely acknowledged to be a major source of economic growth and competitive advantage [5, 6]. New product and process ideas originate from a variety of sources [7]. These innovation sources can be categorized as originating as a result of efforts substantially from within the firm, and those that are accessed external to the firm and transferred in to it [8]. The value of each source depends on the firm's existing stock of knowledge and its ability to access, absorb and exploit new ideas [9, 10]. Previous research has identified several sources of innovation, including users, customers, consumers, suppliers, external scientific expertise, internal R&D, universities, and competitors [8, 11-17]. The insights from lead users are particularly emphasized as a source of novel product concepts in marketing research [18].

The transition of business from CM to OBM inherently changes the importance of each source of innovation. Some of the sources used for a CM-based business might be no longer available or useful to an OBM business. New sources might need to be discovered so as to generate the right ideas in response to market demands. Managing the change in the sources of innovation appears to be of vital importance for the firm's survival and sustainable advantage [13].

Innovation Capability

Innovation is the mechanism by which organizations produce new products, processes and systems required for adapting to changing markets, technologies and modes of competition [11, 19]. The sources of innovation can be defined as where the ideas originate from. However, a firm may not be able to exploit all sources with a particular configuration of resources and capability [20]. Lawson & Samson [13] regard innovation capability as an integration capability to mould and manage multiple capabilities, and to continuously transform knowledge and ideas into new products, processes, and systems for the benefit of the firm and its stakeholders [10, 21, 22].

Firms in transition require both inside-out and outside-in capabilities [23]. The inside-out capability can produce low cost products on a timely basis while the outside-in capability can understand the evolving requirements of customers and energize the organization so as to respond to them. Day [23] also emphasized that a market-driven organization should have superior marketing sensing, customer linking, and channel bonding capabilities. These are the distinctive capabilities required to outperform the competition.

Innovation Alliance

Firms can obtain the desired technology, technological know-how, and skills through a combination of internal and external sources [24, 25]. These may include in-house R&D, joint venture, corporate mergers and acquisition, licensing and purchasing.

Prior research has generally classified the governance structure of inter-firm alliances in terms of their hierarchical components and has typically differentiated alliances by the presence or absence of equity [26-28]. Equity alliances include any exchange agreement in which the partners share or exchange equity. Equity has been considered to be an effective mechanism for managing the rent appropriation concerns associated with partnering. Equity alliances include merger and acquisition, joint venture, and minority investment. Contractual alliances or non-equity based alliances do not involve the sharing or exchange of equity, nor do they entail the creation of new organizational entities [28]. Under the resource constraint and time-to-market pressure, alliance formation could be an effective means to overcome the capability gap as well as to access new sources of innovation.

Resource-based view

The resource-based view (RBV) characterizes the firm as a collection of resources and capabilities rather than a set of product-market positions [29]. The resource-based view proposes that sustained competitive advantage is derived from possessing a set of unique resources that create value in the marketplace. According to Barney, a resource is valuable when it enables strategies that improve efficiency and effectiveness (i.e., exploit opportunities and /or neutralize threats) [30]. Uniqueness derives from being rare (at most, only a few firms have the resource), having imperfect imitability and non-substitutability [31]. Such resources may be physical, such as product designs and production techniques, or intangible, such as brand equity. Teece et al. [32] also speak of the capabilities being honed to a user need, unique and difficult to replicate. Uniqueness and value are both necessary, but neither is sufficient for competitive advantage.

Dynamic capabilities theory emphasizes that competitive advantage rests on distinctive processes, shaped by the firm's unique asset and paths followed. Dynamic capabilities also emphasize the importance of management capabilities and inimitable combinations of resources that cut across all functions, including R&D, product and process development, manufacturing, human resources and organizational learning [13, 33].

Due to the different innovation trajectories from those in the developed countries, existing resource-based perspectives may not be adequate for dealing with the innovation positions, paths, and processes [34]. It may make sense to use dynamic capabilities to build new resource configurations and move into fresh competitive positions without maintaining path dependence. The path breaking change (i.e. CM to OBM transition) may occur in cases where expansion incentives and competitive pressures outweigh path dependence [35, 36].

Lavie [37] proposes how capability can be reconfigured. In his study, he identifies three mechanisms: (1) capability substitution, which offers an immediate response at the level of the overall capability portfolio; (2) capability evolution, which involves continuous experimentation that occurs at the level of particular routines; and (3) capability transformation, which is an intermediate response that applies at the level of a particular capability. The nature of business and technological change, and the capability in question, affects the choice of reconfiguration mechanism, which in turn enables the firm to achieve the desired capability configuration.

Synthesis

The literature reviewed provides a foundation to explore the key factors during the CM-OBM business transition. The change of sources of innovation forces firms to acquire new capabilities through a variety of means. The factors that influence a firm to choose between internal development and external collaboration are critical to firms in transition. Internal development may give better governance over the innovation output while external collaboration can ease the cost to research and development. With the transition toward OBM business, the firms pursue a 'path-breaking' strategy (i.e., one that is substantially different from their current one, and one which is not shaped by its past activities). Either through internal development or external innovation alliance, the capability would be reconfigured by substitution, transformation, and evolution. Based on the literature, a conceptual framework is proposed as an initial point for the following research. It also provides the guideline for data collection. The conceptual framework is illustrated as Fig. 2.

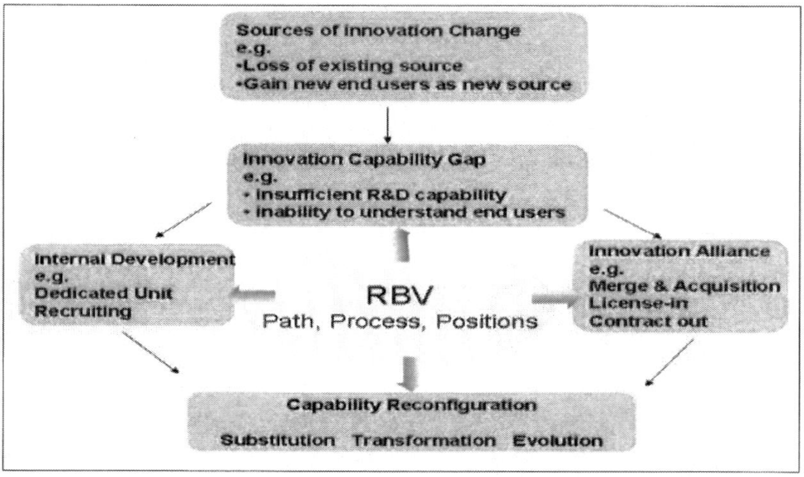

Figure 2. Conceptual framework

Research Methodology

Understanding how innovation capabilities can be reconfigured in the transition from CM to OBM requires access to information that may be commercially sensitive and confidential. The lack of existing public knowledge of the subject and the requirement of the degree of access necessary to obtain the data means case studies are the most appropriate method [38].

The case presented in this paper is one of a series of detailed case studies of Taiwanese computer companies undertaken for program of PhD research. Each of these cases was developed using a combination of analysis of secondary material and semi-structured interviews with senior members of the management team responsible for key business units during the CM-OBM transitions.

In this paper, we are presenting the case of Acer, a leading Taiwanese PC firms which has experienced major transitions in its 32 year history, and has completed a cycle of transitions from

OBM to CM to OBM. The case data is drawn from analysis of substantial secondary sources supported by interviews with several senior managers including the company co-founder, the CEO and the CTO.

Case Study: The Birth of Acer

Founded in 1976 by Stan Shih, Acer (initially named 'Multitech') was the first Taiwanese computer company to try to build its own brand name in foreign markets. It is now Taiwan's largest brand name computer exporter and one of the top three personal computer brand names in worldwide PC markets [39].

The period 1976-1980 marks Acer's start-up phase. The initial product lines were located in Taiwan, where cheap labor and low cost industrial facilities were available. Acer moved into the Hsinchu Science Park near Taipei in 1981. The company started making clones of the IBM personal computer shortly after it was introduced in 1981[1]. IBM-based computers, peripherals, and accessories, manufactured for other companies, became Multitech's largest product lines.

The Dragon Dream

As the company approached its tenth anniversary, it announced a plan for the next ten years that was described as the "Dragon Dream" plan. With the revenues of US$150 million in 1986, the goal was to achieve sales of US$5 billion by 1996. In several Asia countries, the company was already a major player. For example, in Singapore, it had a 25% market share in 1986. The Dragon Dream was built on the Asian base, and the newly offices in Europe and the United States led Stan Shih to create the slogan, "The Rampaging Dragon Goes International." To implement the plan, the need to identify potential overseas acquisitions, to set up offshore companies, and to seek foreign partners and distributors was emphasized.

---

[1] Interview with co-founder, Fred Lin, 16/4/2007

The Vision for the Year 2000

At Acer's 1992 international distributors meeting in Mexico, the company articulated a commitment to linking the company more close with its national markets, describing the vision as Global Brand, Local Touch [40]. It envisioned Acer evolution from a Taiwanese company with offshore sales to a truly global organization with deeply-planted local roots. To fulfill the vision for the year 2000, Stan Shih and his management team developed the strategy consisting of three elements:

1) Joint ventures with partners overseas;
2) "Fast food" style manufacturing and assembly operations; and
3) "Client-server" organization structure [41].

One key factor in the strategy "Global Brand, Local Touch" vision was to form joint ventures with partners in foreign markets where the partners would take majority interests. Doing so could allow Acer expand into overseas markets without requiring significant capital investment. The financial success and risks were shared between Acer and its partners.

The Fast Food business concept

In May 1992, Compaq announced a 30% across-the-board price reduction in the price of its PCs. Industry expectations were for a major shakeout of marginal players.

Taiwanese PC manufacturers traditionally shipped PCs by sea to their intended overseas markets. Several problems coexisted with this arrangement. It usually took an average of four to five weeks in total to manufacture components and build PCs in Taiwan, then ship by sea to the United States, go through import channels and finally reach the customer. Because of the falling prices and currency fluctuation, by the time PCs finally reach store shelves, they were often overpriced. Secondly, dealers and distributors must accurately forecast what configurations (e.g., microprocessor, amount of memory, amount of disk storage, type and number of

I/O slots, etc.) customers will buy or else keep a large inventory of PCs of different configurations to satisfy demand. Thirdly, PCs had become commodities, and retail channels increasingly asked for slightly different configuration of their products which may be difficult to accomplish from several thousand miles away.
To overcome these problems, Acer began to create new initiatives. The System PC unit developed the "ChipUP" concept. This patented technology allowed a motherboard to accept different types of CPU chip—various versions of Intel's 386 and 486 chips—drastically reducing inventory of both chips and motherboards. Another unit, Home Office Automation, developed the "2-3-1 system" to reduce the new product introduction process to two months for development, three months for selling and one month for phase-out.

Acer was not the first to apply distributed manufacturing concept to the manufacture of PCs. Some PC manufacturers in Taiwan had already discovered that in order to compete, they had to source casings and some other components locally and assemble locally, but they did not approach this systematically. Acer adapted the concept to fit its environment and further refined it. In order to fully implement this manufacturing strategy, Acer registered intellectual property relating to its PCs for modular assembly as well as configuration flexibility. As a result, Acer had four to five designs of motherboards and three chassis. By the patented technology of "ChipUp", every motherboard could accommodate different Intel microprocessors and any motherboard could fit into any chassis. This flexibility also allowed each assembly site to put together PCs with the exact microprocessor model, memory, hard disk storage that was appropriate for its market.

This "fast-food" model allows Acer to maintain control over quality, keep inventory to a minimum, and assemble competitively priced "fresh" PCs on a made-to-order basis. Between 1991 and 1993, the implementation of this strategy resulted the amount of Acer's inventory falling from 90 days to 45 days [41].

The shift to distributed manufacturing was complemented by an added emphasis on OEM sales of components. OEM sales accounted for 32% of the computer products business revenue in 1993 and accounted for over 50% in 1997. OEM/ODM business became increasingly substantial in Acer's revenue. (See Fig. 3).

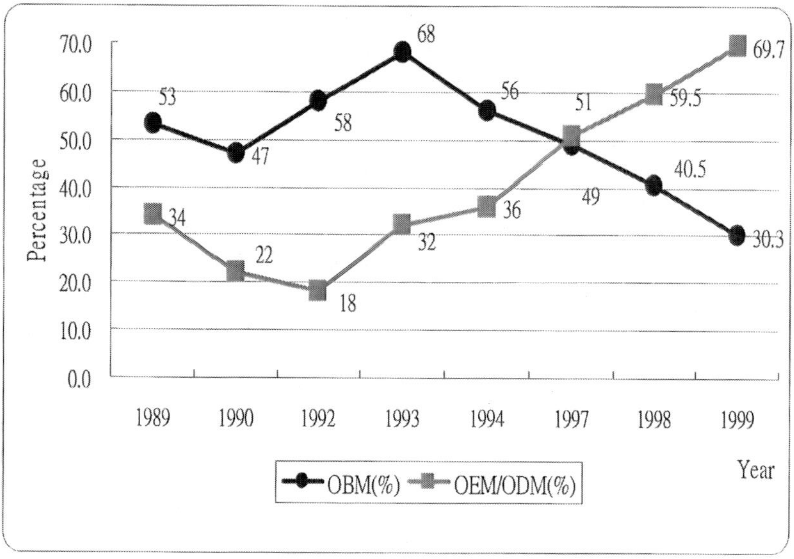

Figure 3. Acer's OBM and OEM revenue proportion of selected years, Source: data compiled from various annual reports

Client Server Organization Structure

To reinforce the more networked approach of this new strategy, Acer introduced its "client-server" organization model in 1993[2]. By using the metaphor of the network computer, the role of the Taiwan headquarter as a "server" uses its resources (such as finance, human resource, intellectual property etc.) to support its "client" business units that controls key operating activities. Under this organizational model, business units could leverage their own ideas or initiatives directly through other RBUs or SBUs without having to go through the corporate centre. Corporate headquarter functioned to help and mediate rather than to dictate or control.

[2] Acer website. http://global.acer.com/about/milestones.htm

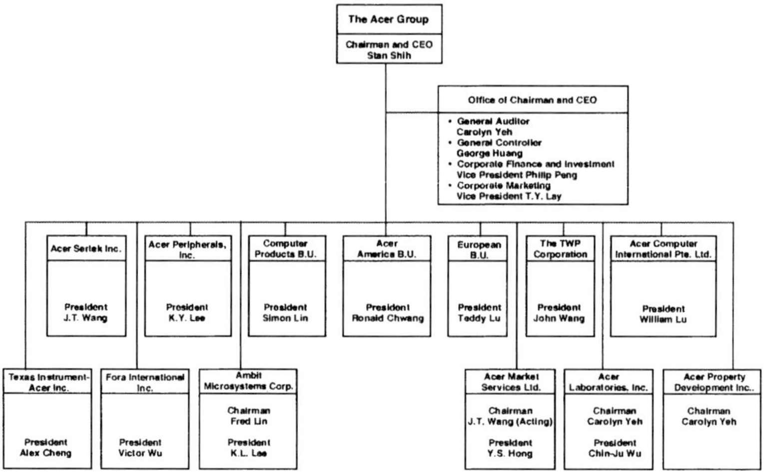

Figure 4. The Client-Server organization

Acer believed that this model would allow the organization to develop speed and flexibility as competitive advantages. (See Fig. 4).

Acer's Challenge in 2001

Acer faced several challenges in 2001. The global economic slowdown and the bursting of the internet bubble precipitated a global decline in PC demand starting in late 2000. Companies started cutting costs on areas from training and IT budgets to sales and marketing, and that lead to further decrease in sales.

Acer made good use of the low labor cost in China as its most important production location. However, it had never had established itself as a leading brand in developed markets such as the United States. In 2000, Acer only managed to make a net profit out of selling investment and share in subsidiaries. Selling inventory to overseas units might have increased the market share of Acer brand products but at the expense of profit. An executive of a competing motherboard company commented in 1998: "to move from ODM to OBM with all the necessary logistical support is tough. IBM and Compaq do not do their own manufacturing. But Acer manufactures its own components. It's all overhead burdens." [42]

Acer ran into issues combining CM and OBM businesses in parallel. According to one analyst, "The key complaint from customers as well as competitors has been that Acer has done both own brand business as well as design and manufacturing services business. It has caused a lot of conflict of interests" [39, 43]. Since Acer might manufacture a PC model for a client and a similar Acer-brand one in direct competition, CM clients feared leakage of business secrets. Acer's CM business suffered after IBM, a major customer, cancelled an order in 2000. IBM accounted for 53% of Acer's total revenue in the first quarter of 2000, but only 26% in the second quarter of 2001[3].

The Way to OBM

The Acer Group announced the first steps in a major reorganization in December 2000[4]. The new Group would consist of three divisions: Acer Brand Operation (ABO), Design Manufacturing Services (DMS), and Hold and Investment Business (HIB) (See Figs. 5 and 6). The reorganization would reserve the decentralization of the 1998 restructuring to avoid competition among subsidiaries and promote internal communication.

In May 2001, Acer announced it would separate ABO and DMS by spinning off the latter into a new company, Wistron Corporation. ABO, which would become the flagship business of the Acer Group, would concentrate on marketing and sales for Acers own brand products. It would outsource all its manufacturing and would give a free hand to its contract

[3] Interview with Fred Lin, 16/04/2007

[4] Acer website. http://global.acer.com/about/milestones.htm

manufacturers. Plans were to contract out work to the lowest bidder, which might not be within Acer Group. J. T. Wang would be the CEO of ABO. ABO would cut down the variety of products sold in countries where Acer was relative weak. Acer used to have hundreds of different PC models on sale globally, but after the organization it would sell only six different own-brand desktop models and three or four notebook models. It was hoped that these moves would cut down inventory size and marketing costs, and enhance economies of scale.

Wistron's initial capital was US$30 million, but another US$280 million would be injected by the Acer Group. With eight global operation sites, the DMS claimed to have achieved US$3 billion in OEM business revenue in 2000 and to be the top global ODM/OEM supplier for the PC industry in 2001.[5] The spin-off

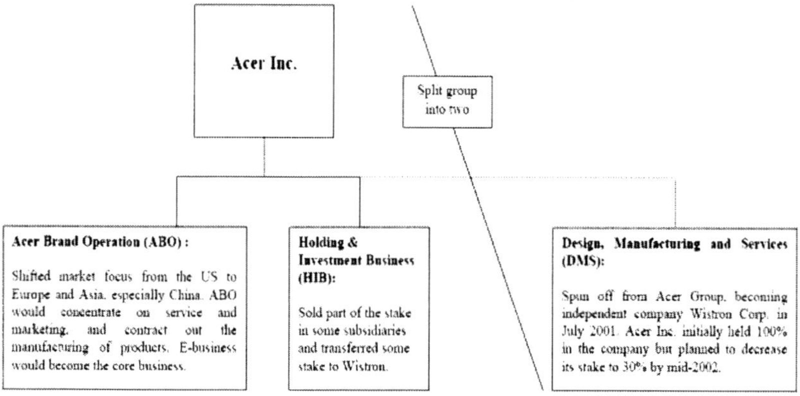

Figure 5.Acer's reorganization structure in 2001
Source: Burns, 2001; Press release, Wistron Corporation, 10 July, 2001; and Press release, Acer Inc, 22 June, 2001.

was expected to allow Wistron to make the best use of its research and development resources, design capabilities, and global logistics system, and to have a free hand in looking for new customers. At its inception, Wistron would have clients that include IBM, Dell, Hitachi, and Fujitsu, with IBM the single largest client.

[5] Press release, Wistron Corp., 10 July, 2001.

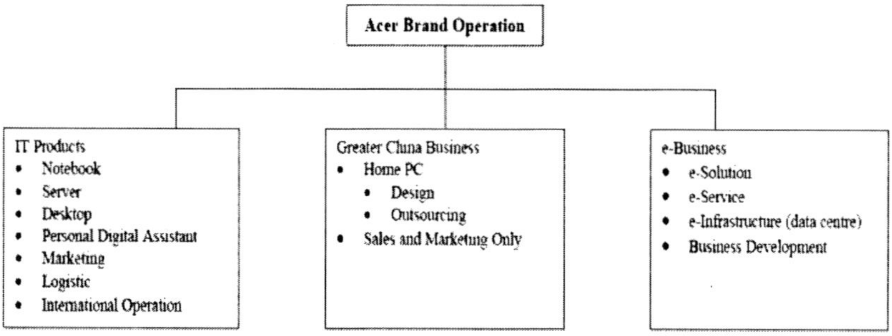

Figure 6. Business divisions of Acer brand operation after the 2001 reorganization, Source: Company data as cited by Credit Suisse First Boston (Hong Kong) Limited, 26 July, 2001

The reorganization was backed by an important change in business concept. Acer's main revenue sources had been the manufacturing of PCs and motherboards. The new Acer would shift to the service oriented business model from a traditional hardware business[6]. Acer hoped that e-business would become a major part of ABO and would represent the new direction of the Acer Group as an internet enabler. ABO would provide e-service and e-solutions as value added services to customers. Taiwan and Mainland China are target markets in the service business as beginning. Other markets would be considered afterwards.

Table 1. The new Acer vs. the old Acer, Source: company data

	New Acer	**Old Acer**
Industry focus	Service	Manufacturing
Era	Knowledge-based economy	Information technology
Corporate culture	User-centric	Technological innovation
Core competence	Marketing and service	IT products
Mission	Pioneer	Pioneer

[6] Interview with RC Chang, CTO, 18/04/2007.

The Birth of Value Lab, Acer's New R&D Centre

Acer's original R&D lab was established in the late 1970s to conduct hardware R&D, and by the mid-1980s, it had developed the capacity to conduct R&D for software products. The primary objectives were the commercial development of all of Acer's products and hardware components, which mainly had been transferred from the U.S. as prototypes. This involved integration with hardware or other components, product testing, ergonomic design, and customization for regional markets.

Many of these projects required highly customized efforts, and they would not be transferable to any other parties. Thus revenue was generated through license royalties as well as non-return engineering charges, which involved customer pre-paying for work done on a project by project basis that was non-refundable if the client chose to discontinue the project due to a change of business strategy or products focus.

To increase its competitiveness and access to knowledge, Acer also forged strategic joint venture with other Taiwanese software developers to enter the niche markets such as writing software for OEM manufacturers and embedded systems for imported hardware into the Taiwanese market.

Acer set up an R&D site in the technology hub of Silicon Valley in the late 1980s. The mission of this site was to develop prototypes, and also taking responsibility for identifying many of Acer's strategic partners. The relationships were formed partially with a view to establishing joint ventures with local partners, s strategy to which Acer was highly committed.

The "Value Lab" was established according to the change of business scope and concept. In the past, most of the R&D activities focused on the product development following the standards set by dominant companies such as Intel. The mission of R&D under the CM-focused business model was to improve quality, efficiency, and cost. The Value Lab comprised two centers, which are product innovation centre and service value innovation centre.

For the short term objectives, the Value Lab needed to meet the demands and requirement from internal business units. The product manager from business units would cooperate with the Value Lab to launch new product in the period of 3 to 6 months. For the middle term objective, the mission was to understand customers' needs. The Value Lab not only conducted global marketing research on regular basis but also cooperated with consulting firms for specific projects. The Value Lab also formed collaborations with partner companies for next generation products or technologies. Acer's CTO[7] emphasized that the Value Lab's task was to integrate the state of the art technology into products, identify new applications and create value accordingly. The long term objective was set on precompetitive technology and early stage research and development in the areas that it is unlikely to be profitable in the short term[8].

With comparatively limited expense in R&D, the Value Lab also cooperated intensively with academic and research institutes. The Advanced Technology Centre in Industrial Technology Research Institute (ITRI), Institute for Information Industry, and Massachusetts Institute of Technology (MIT) are all partners with on-going projects respectively. For example, Oxygen Alliance[9] is a US$50 million five-year project, launched in June 2000, that intended to increase computer use and make their operation more human, creating technologies to replace desktop computers and keyboards with smaller, portable electronic devices [44]. The Value Lab is the Acer's arm in the Oxygen Project.

---

[7] Dr. RC Chang, formerly a professor in National Chiao-Tung University.
[8] For example, solutions for voice recognition, image, and human interface are under developed in the market. No existing solution is available. This is where the Value Lab would explore.
[9] More details at the official website, http://oxygen.csail.mit.edu/Overview.html

Discussion

Acer's dramatic change in 2001 triggered by the crisis has created a new era for Acer. After twenty years of manufacturing, it became a company with emphasis on R&D, marketing and sales and thus "manufacturingless". The innovation capability transformed from engineering and manufacturing to marketing. R&D activities not only involved the field of technology but also the consumer behavior, usage and attitudes. In addition to the internal development of required innovation capability, the collaboration with external institutes also provided Acer with considerable breadth of innovation capability, at a relatively low level of investment in technical research and development.

When Acer's focus was on the CM business, its innovation capability came from its capacity to absorb new technological knowledge from its clients. When the CM business evolved into ODM business, Acer made significant achievement in two fields. The growing innovation capability allowed Acer to introduce new products of equivalent performance to the first tier PC companies such as IBM. Acer also managed to innovate via its novel approach to distributed manufacturing which made Acer more competitive in the CM/ODM business.

When Acer decided to transform into OBM business, organizational re-engineering was conducted. After spinning off its manufacturing arms, Acer was relieved from the conflicts with their CM clients and free to look for the best ODM service from the market. The establishment of the Value Lab and the implementation of an open innovation strategy complemented the constrained resources Acer could afford to invest in technological research. The collaboration with external institutes enhanced the technological innovation capability while the regular survey and research about consumer educated Acer about the end users – something that had been less important under a CM business model.

Conclusion

The objective of this paper has been to explore how a Taiwanese PC firm has reconfigured its innovation capability to allow it to transition from CM/ODM to OBM. The case of Acer has shown the type of challenges faced in moving between CM and OBM, and potential means of overcoming them. The Acer case has illustrated how a combination of internal and external approaches was used to allow the firm to reconfigure innovation capabilities in a way that did not require substantial capital investment in internal R&D. The conceptual framework has allowed us to examine the substitution, transformation and evolutionary aspects of Acer's approach.

Work is currently underway to apply this framework to two further case studies from within the Taiwanese computer industry: BenQ and Asus. From these three cases in combination, it is anticipated that useful lessons can be drawn to assist firms in managing transitions from CM to OBM business models.

References

1. Ernst, D., *Late innovation strategies in Asian electronics industries: a conceptual framework and illustrative evidence*. 2004, East-West Center. p. 28.
2. Kim, L., *Imitation to Innovation: the dynamics of Korea's Technological Learning*. 1997, Boston, MA: Harvard Business School Press.
3. Hobday, M., *East Asian latecomer firms: learning the technology of electronics*. World Development, 1995. **23**(7): p. 111-1193.
4. Hobday, M., *Taiwan: small firm innovation clusters*, in *Innovation in East Asia: the challenge to Japan*. 1997, Edward Elgar: Cheltenham. p. 95-135.
5. Ernst, D., *Pathways to innovation in the global network economy*. 2003, East-West Center. p. 46.
6. Schumpeter, J., *Capitalism, Socialism and Democracy*. 1942, New York: Harper Brothers.
7. Montalvo, C., *What triggers change and innovation?* Technovation, 2006. **26**: p. 312-323.
8. Bommer, M. and D. Jalajas, *Innovation sources of large and small technology-based firms*. IEEE Transactions on Engineering Management, 2004. **51**(1): p. 13-18.
9. Rosegger, G., *Firms' information sources and the technology life cycle*. Int. J. Technology Management, 1996. **12**: p. 704-716.
10. Cohen, W.M. and D.A. Levinthal, *Absorptive capacity: a new perspective on learning and innovation*. Administrative Science Quarterly, 1990. **35**: p. 128-152.
11. Dougherty, D. and C. Hardy, *Sustained product innovation in large, mature organizations: overcoming innovation-to-organization problems*. Academy of Management Journal, 1996. **39**(5): p. 1120.
12. Gerybadze, A. and G. Reger, *Globalization of R&D: recent changes in the management of innovaiton in transnational corporations*. Research Policy, 1999. **28**: p. 251-274.

13. Lawson, B. and D. Samson, *Developing innovation capability in organizations.* International Journal of Innovation Management, 2001. **5**(3): p. 377-400.
14. Sher, P.J. and P.Y. Yang, *The effects of innovative capabilities and R&D clustering of firm performance: the evidence of Taiwans semiconductor industry.* Technovation, 2005. **25**: p. 33-43.
15. Veugelers, R., *Make and buy in innovation strategies: evidence from Belgian manufacturing firms.* Research Policy, 1999. **28**: p. 63-80.
16. von Hippel, E., *The sources of innovation.* 1995, New York: Oxford University Press. 218.
17. Laursen, K. and A. Salter, *Searching high and low: what types of firms use universities as a source of innovation.* Research Policy, 2004. **33**: p. 1201-1215.
18. von Hippel, E., *Lead users: a source of novel product concept.* Management Science, 1986. **32**(7): p. 791-805.
19. Utterback, J.M., *Mastering the dynamics of innovation: how companies can seize opportunities in the face of technological change.* 1994, Boston, MA: Harvard Business School Press.
20. Gregory, M.J., *Technology management: a process approach.* Journal of Engineering Manufacture, 1995: p. 347-356.
21. Henderson, R.M. and K.B. Clark, *Architectural innovation: the re-configuration of existing product technologies and the failure of established firms.* Administrative Science Quarterly, 1990. **35**(1): p. 9-31.
22. Kanter, R.M., *Swimming in newstreams: mastering innovation dilemmas.* California Management Review, 1989: p. 45-69.
23. Day, G.S., *The capabilities of market driven organizations.* Journal of Marketing, 1994. **58**: p. 37-52.
24. Ford, D. and D. Farmer, *Make or buy -- a key strategic issue.* Long Range Planning, 1986. **19**(5): p. 54-62.
25. Granstrand, O., *Toward a theory of the technology-based firm.* Research Policy, 1998. **27**: p. 465-489.

26. Pisano, G., *The governance of innovation: vertical integration and collaborative arrangement in the biotechnology industry.* Research Policy, 1991. **20**: p. 237-249.
27. Pisano, G., *The R&D boundaries of the firm: an empirical study.* Administrative Science Quarterly, 1990. **35**: p. 153-176.
28. Gulati, R. and H. Singh, *The architecture of cooperation: managing coordination costs and appropriation concerns in strategic alliances.* Administrative Science Quarterly, 1998. **43**: p. 781-814.
29. Wernerfelt, B., *A resource-based view of the firm.* Strategic Management Journal, 1984. **5**(2): p. 171-180.
30. Barney, J., M. Wright, and D.J.J. Ketchen, *The resource-based view of the firm: Ten years after 1991.* Journal of Management, 2001. **27**: p. 625-641.
31. Medcof, J., *The resource-based view and transnational technology strategy.* Journal of High Technology Management Research, 2000. **11**(1): p. 59-74.
32. Teece, D.J., G. Pisano, and A. Sheun, *Dynamic capabilities and strategic management.* Strategic Management Journal, 1997. **18**(7): p. 509-533.
33. Teece, D.J. and G. Pisano, *The dynamic capabilities of firms: an introduction.* Industrial and Corporate Change, 1994. **3**(3): p. 537-556.
34. Hobday, M., *Firm-level innovation models: perspectives on research in developed and developing countries.* Technology Analysis & Strategic Management, 2005. **17**(2): p. 121-146.
35. Eisenhardt, K.M. and J.A. Martin, *Dynamic capabilities: what are they.* Strategic Management Journal, 2000. **21**: p. 1105-1121.
36. Karim, S. and W. Mitchell, *Path-dependent and path-breaking change: reconfiguring business resources following acquisitions in the U.S. medical sector, 1978-1995.* Strategic Management Journal, 2000. **21**: p. 1061-1081.

37. Lavie, D., *Capability reconfiguration: an analysis of incumbent responses to technological change.* Academy of Management Journal, 2006. **31**(1): p. 153-174.
38. Yin, R.K., *Case study research: design and methods.* Applied Social Research Methods 2003, Thousand Oaks, California: Sage Publications.
39. Lu, Y.W., *Cover story*, in *Business Next*. 2006.
40. Acer, *Annual Report*. 1993, Acer Inc.: Taipei.
41. Shih, S., *Me Too is Not My Style.* 1997, Taipei: CommonWealth Publication Inc.
42. Kovar, J., in *Computer Reseller News*. 1998.
43. Burns, S., *Cover story*, in *Far Eastern Economic Review*. 2001.
44. Heaney, B., *MIT shares artificial intelligence information*, in *Taipei Times*. 2003: Taipei.

The Emergence of Wireless Networks

Wenshin Chen

ABU DHABI UNIVERSITY, UAE

David Bennett

Aston University, UK
and
University of South Australia, Australia

The purpose of this chapter is to provide organisations with a better understanding of how they can manage rapidly developing information and communication technology, and particularly wireless networks that have evolved into several technological standards in recent years. Case analyses of two large organisations, called here Metro University and Urban Healthcare Centre, demonstrated collective patterns of how organisational legitimacy was changed. More specifically organisational legitimacy, in relation to wireless networks and IT management, was often enhanced by the recognition and/or appreciation from three major groups of stakeholders: external customers who were normally prospective users, internal customers who were the existing users, and the IT staff members who provided services to the user groups. Our empirical findings suggest that certain social values such as prospective customers' recognition, internal customers' satisfaction, and existing IT staff members' appreciation might help shape an organisation's long turn image and legitimacy. They might not provide immediate profitability and growth, but their effects could be sustained over time and help organisations to survive and ultimately thrive. Organisations situated in today's rapidly changing IT and wireless environment might thus need to reconsider how to better capture various social values derived from IT and wireless network projects, and in turn help legitimise their innovative status that might ultimately ensure long-lasting survival.

***Keywords*: wireless networks, ICT, legitimacy, case study, social value.**

Background

Information and communication technology (ICT) has captured increasing attention in industry and academia over the years. One of the fastest growing examples is that of wireless networks and related mobile devices. For instance *The Economist* reported a series of cover stories in April 2007, all related to wireless and mobile phenomena. The topics of those stories included wireless chips and smart services, sensor networks and smart devices, various forms of emerging wireless communications, ubiquitous apartments where everything (i.e. home appliances, consumer electronics, and mobile communications) was controlled and connected online, growing communications chips that would soon vanish existing wireless tools, and a seemingly RFID (Radio Frequency Identification) tag injected into a club patron's arms for entry and purchasing records (Economist, 2007a). A variety of issues were also discussed including wireless energy and invisible security and privacy issues (Economist, 2007b). *The Economist* cover stories represented a continuously growing attention to mobile phenomena in recent years.

The emergence and rapid development of these technologies created a new channel for business models and core competition. In education, mobile technology facilitated learning process by enhancing students' interests in their individual works (Patterson, 2001); some private schools have successfully launched laptop programmes to facilitate the notion of anytime, everywhere learning (Rajala, 2003). In the health care industry, wireless technology has empowered physicians and enhanced their productivity and patient care (Fiser, 2004); many medical schools have required students to have a PDA (Personal Digital Assistant) prior to entering the programmes (Rege & Kean, 2003). In agriculture, wireless local area networks (WLANs) have been successfully tested to allow rapid data flow among all farming equipments and in turn resolved bottleneck issues for precision farming (McKinion et al., 2004). In the oil and energy field, recently emerging WiMAX (Worldwide Interoperability for Microwave Access) networks has been launched in Yemen

(Hoskins, 2007). In the government sector, one of the electronic information sources most recently indicated that a number of cities in the US would soon have region or city wide wireless networks implemented (Muniwireless.com, 2006). In the U.K., the River Thames in London is being transformed into an enormous hotspot allowing wireless connectivity on the river and along its banks (McCue, 2007).

Situated in such competitive environments, organisations will inevitably face managerial issues of these rapidly emerging ICTs and seek better approaches to make appropriate decisions and justifiable investment. The purpose of the research investigation on which this chapter is based thus seeks to help organisations better understand and manage these mushrooming ICTs, particularly in relation to wireless networks that have quickly evolved into several technological standards in recent years. Our particular interests focus on social dimension of IT values that are largely overlooked in the existing literature as shown in the section that follows.

Literature Review

In general, wireless networks consist of many emerging and complex components including wireless personal area networks (WPANs), wireless local area networks (WLANs), and wireless metropolitan area networks (WMANs) (standards.ieee.org). The architecture or standard for each wireless network service is so dynamic (Varshney & Vetter, 2000) that it is most feasible to focus on one specific component in one study. This study thus concentrates on the implementation of wireless local area networks (WLANs) and the usage of wireless devices connecting to the networks. In addition, various WLAN standards have evolved over time. Most notable ones include 802.11b, 802.11a, 80211g and the emerging 802.11n. The development timeline of 802.11b and 802.11a was approximately in 1999 while that of 802.11g was in 2003. Although 802.11g is considered the most commercially popular standard in the current market, a more recent development, 802.11n, that possesses nearly five times greater transmission rate than that of 802.11g and twenty times more than that of 802.11b, is estimated to be commercially available in less than two years (standards.ieee.org). In other words, almost every other year a more capable wireless network standard has emerged and replaced the exiting one. In a larger organisation where wireless networks might be multifarious and the implementation process could take years to complete, this rapid development of wireless network standards naturally complicate organisations' choices of wireless network technology and their realization of its values.

In the current literature, how an organisation evaluates information technology and in turn understands its strategic values is often based on economic perspectives. Some typical examples include the measures of productivity (Hitt & Brynjolfsson, 1996) and efficiency (Banker, Kauffman, & Morey, 1990). Information technology is often perceived to serve the purpose of enhancing operations efficiency and productivity and subsequently reduces production or operations cost and/or increases profitability. Although these empirical research studies provide substantial evidence to support valuable arguments, they tend to simplify strategic values of information technology because the essential social perspectives of IT values (Smithson & Hirschheim, 1998) have been largely overlooked (Christiaanse & Huigen, 1997). While economic perspectives emphasise monetary gains and quantifiably measures, social perspectives allow intangible values to surface that may not be quantifiable. These non-quantifiable social values provide alternative understanding of how IT might shape an organisation's long term survival, particularly when the greater technological environment is highly uncertain and constantly evolving such as the wireless network developments faced by organisations today.

A. Legitimacy and IT

One of these non-quantifiable and largely overlooked social values is an organization's legitimacy. As institutional theorists argue, organisations often act to gain legitimacy and the social and cultural fitness of the field, not to increase cost benefits or efficiency (DiMaggio & Powell, 1983). Many of the ideas, values, and beliefs that an organisation acts upon are influenced by institutional forces (Meyer & Rowan, 1977). To survive, an organisations has first to obtain institutional recognition; acting to accommodate institutional expectations, regardless of efficiency or cost benefits, would enhance institutional recognition and, in turn, help organisations to compete and survive (Lawrence, 1999; Oliver, 1991; Scott, 1987). In modern society where internal and external organisational issues are rather complex and uncertain, organisations tend to follow the standardised practices or procedures that are highly sanctioned in the field (Slack & Hinings, 1994). Failing to follow highly recognised social activities means that organisations do not meet the basic requirement to compete, and consequently will not be recognised as a legitimate player in the market (Dirsmith, Fogarty, & Gupta, 2000). Thus, the possibility for them to survive will be limited.

In the IT literature, legitimacy issues in organisations have gained increasing attention (Christiaanse et al., 1997; Jarvenpaa & Leidner, 1998; King et al., 1994). For instance, King et al. (1994) discussed the institutional intervention - influences and regulations inserted by various institutions - in IT innovation. Christiaanse and Huigen (1997) investigated institutional dimensions in IT implementation and suggested that cultural biases and structural order, along with economic driving forces, would influence the implementation of inter-organisational systems. Jarvenpaa and Leidner (1998) studied the cultural and environmental elements of institutional influences in the context of a developing country and discussed broadly how overcoming cultural and environmental challenges could lead to better development of an IT company. Due to the emerging nature of wireless network technology little empirical research has paid attention to its relationship with an organisation's legitimacy. Our research endeavour thus extends the existing knowledge of the above mentioned studies and further seeks to unveil how wireless networks shape and reshape an organisation's legitimacy in its competitive marketplace and uncertain technological environment. The underlying research question that guides our investigation is, "What are the social values gained from the emergence of wireless networks in organisations?" More specifically, "How is organisational legitimacy enhanced through the implementation of wireless networks?"

Methodology

We employed the case study research methodology to understand more deeply each participant organisation's experiences of wireless networks. In contrast to most quantitative studies aiming for generalisation, case study research is most suitable for obtaining a deep understanding of situations within single settings (Eisenhardt, 1989) and is particularly appropriate for exploratory studies (Yin, 1994). Since empirical research regarding wireless networks in general, and its relationship with organisational legitimacy in particular, has rarely been examined,the case study methodology serves the exploratory purpose of this research investigation.

B. Case study sites

Two case organisations, one from the higher education sector and the other from the health care sector, were selected because of these sectors' substantial investments in advanced technology in general, and wireless networks in particular. The case organisations were also two large well-known, not-for-profit organisations within the same metropolitan area in the U.S. Their leading status in the area and complex organisational structure would naturally shape an interesting wireless implementation process and in turn provide valuable insights for other organisations in the field. Moreover their IT systems, one decentralised and the other centralised, could further contrast the wireless implementation process and results and shape different realisation of IT values, including non-quantifiable social dimensions on which this research investigation focuses.

The first case organisation, Metro University[10], is a large public research university with more than 35,000 students coming from immensely diverse backgrounds. While the organisation is not necessarily best known for its technological innovation, its distance education and related technologies well meet many commuting students' needs and it is consequently considered to be one of the leading organisations in the state. Metro University's IT structure is mostly decentralised due to its large physical size and student population.

The second organisation, Urban Healthcare Centre, is a renowned hospital both regionally and nationally housing more than 4000 employees and almost 1000 beds. It is a frequent winner of "Most Wired Hospitals," an annual survey conducted by *HHN (Hospitals and Health Networks)* to identify hospitals that best utilise information technology. In addition, Urban Healthcare Centre is also recognised several times by *Fortunate* as one of the "100 Best Companies to Work For" and by *U.S. News & World Report* as one of "America's Best Hospitals." Its large centralised IT department serves more than thousands of physician and nursing mobile devices.

C. Data collection

Data collection for the research involved primarily semi-structured interviews, which followed the same guidelines, and were supplemented by participant observations and collection of documentation. The research investigation commenced by first contacting each organisation's networking director and subsequent interviews followed soon after. Eventually, twenty three participants in Metro University and eighteen participants in Urban Healthcare Centre were interviewed. Participant observation was also conducted where appropriate. For instance, in Metro University formal permission was obtained several times to observe the professors' classes. Although such participant observations did not generate significant amounts of textual data,

---

[10] All organisations' and individuals' names are pseudonyms in the study.

they helped disclose subtle issues that might not be revealed from formal interviews and thus allowed more precise and authentic interpretation of organisational experiences of wireless networks. Lastly, documents and public records, particularly organisation websites and online information, were also reviewed. This thread of data sources provided significant amounts of information to help understand the basic background of each organisation such as their structure and culture, recent organisational and technological development, and their institutional status in the field. Such information helped make sense of interview information and thus enabled better data triangulation that has frequently been suggested by case study researchers (Eisenhardt, 1989).

D. Data analysis

Data analysis was based largely on textual data derived from digitally-recorded interviews that averaged 55 minutes each. All interview records were transcribed and the resulting transcription comprised 387 single-spaced pages of data. The data analysis techniques were primarily dependent upon analytic coding in general and particularly the use of NVivo text analysis software. The analytic coding strategy that was employed involved two stages: initial coding and focused coding (Lofland & Lofland, 1995). The initial coding was a broad search for any concrete indications that were related to the theoretical lenses; as the quantity of initial codes accumulated, the more frequently used codes were then articulated and in turn, shaped focused codes. The focused coding elaborated more on the frequently used codes and collapsed or dismissed those that were less productive. This strategy was also similar to "pattern coding" (Miles & Huberman, 1984), "pattern-matching" (Yin, 1994), "system closure" (Richards, 2002), or the *Constant Comparative Method (CCM)* (Glaser & Strauss, 1967). Furthermore, the most current version of NVivo text analysis software, NVivo 7, was used to help organise codes and create themes for the social values of wireless networks, particularly in relation to organisational legitimacy.

Results

The case analyses of these two large organisations demonstrated collective patterns of how organisational legitimacy was changed. In Metro University such changes in organisational legitimacy were derived from better recognition by various stakeholders such as prospective students, current students, and staff members. In Urban Healthcare Centre the realisation of legitimacy enhancement mostly stemmed from the higher appreciation of staff members and the general public.

The case of Metro University

As more universities have made attempts to develop wireless campuses, Metro University's implementation of wireless networks provided underlying intangible social values that helped the organisation meet such institutional expectations. Since Metro University's wireless initiatives date from the 1990s the emerging nature of wireless networks has helped legitimise its status internally and externally. Internally, the communication needs across campus justified the central IT department's action to take control of the network implementation and maintenance. Externally, it enabled Metro University to build an improved image and in turn better market itself in the institutional field. As the wireless networks increasingly penetrated the higher education institutions, the external legitimacy that was recognised from the technological advance of wireless networks became more significant in an organisation's institutional competition.

For instance, Ling, a computer science professor, revealed that the contemporary generation who grew up with wireless networks and a variety of multimedia technologies has inevitably taken wireless accessibility in the colleges for granted. When they applied for colleges, the existence of wireless networks might not be the driving factor that affected their decision. It definitely influenced how they perceived the capability of a university.

> But I think that' s probably the trend. Most of the campus in the nation, they are wireless covered... While I was talking to a bunch of high school girls, they are on campus visiting and they are interested in getting to know our computer science programme. One of the questions they asked is whether our campus is wirelessly covered and when I gave her a positive answer, she seemed to be happy. [Ling, Professor]

Internally, the existence of wireless networks in the central library has enhanced its status to being not just an academic resource but also a social centre. With Internet accessibility via wireless networks across the library, students increasingly formed groups to discuss their projects. The availability of laptops in the library for students further encouraged them to utilise every space of the library, which in turn increased the popularity of the library. The traditional role of the library as the academic centre of universities has thus been transformed and strengthened to an all-purpose campus centre.

> We [the library] are very popular. It' s become quite a meeting place. We have people come here to meet their friends. They get together in the group. They like to use our study rooms. They group together. And then the last thing is if they don' t already have their own laptop, they come to us and say, "Let' s get a laptop or 2," and sometimes they are on laptop studying and access on line resourcing··· Sometimes it' s just the matter, I think, some kind of social things as well··· they play the online games, they are watching movies, they are listening to music, they are doing a lot of social actions and take the advantage that we have large building with large wireless network. [Peter, Circulation Manager]

Moreover, the unique communication needs of wireless networks allowed the central IT department at Metro University quickly to regain the control of accessibility and in turn maintain a more robust network across campus. As networking director, Cameron, revealed, there were many unofficial wireless networks in the beginning where individual members established their own access points. In a large environment such as Metro University, the inevitable interference created by scattering access points with unknown standards could very likely generate potential network maintenance issues. Such potential issues legitimately enabled the central IT department to re-centralise its previous decentralised IT structure, at least in relation to wireless networks.

> It' s just such a complex environment that it became a pretty easy sell if I could get up in front of them long enough I could usually convince them that this is the right thing to do··· I' m looking at the long term view. It' s not just buying a switch, or buying a hub or buying an access point, and throwing it into the network and magically it works forever··· it has become very clear you can' t decentralize some sets of services. You still need to maintain at one level a centralized capability moving forward because there is always going to be that newest, latest, greatest technology that you just can' t distribute everywhere. [Cameron, Networking Manager]

As prospective students recognised Metro University's wireless networks, this became a validated factor in their college applications. In the current generation of students' minds Metro University has now become one of the legitimate players in the higher educational field. As students appreciated the wireless environment that Metro University provided, they increasingly transformed its typical academic atmosphere [e.g. the library] into a new social and learning setting as well. Due to its decentralised IT structure, Metro University's central IT department has seized the opportunity that wireless networks provided to legitimise its centralised power and regain its centralised networking management.

1) The case of Urban Healthcare Centre

At Urban Healthcare Centre, the emergence of wireless networks continued to reaffirm its traditional innovative image and long-established medical reputation. As a teaching hospital, Urban Healthcare Centre established wireless networks as a platform to train its prospective nurses and in turn gain greater recognition from their recruiters. For instance, Yvette, a nurse and senior analyst for nursing applications, revealed that Urban Healthcare Centre was her only choice when she considered relocating to the medical centre from another city because its organisational legitimacy in the field was well perceived. Such a perception was prevalent among nursing schools and their students.

> The nurse school that is right there, nurses are using the Centre as a part of their teaching platform, so again they hear a pride to work here so when they come they want to work as graduate nurses. I think it is a big deal. I think it is to their benefit, to nurse recruiters, to have that on their brochures, as a part of their discussion point when they are out recruiting··· [Yvette, Nursing Analyst]

Internally, the same recognition also existed among the Centre's own nursing staff. As nursing administration clearly recognised the benefits of computer systems and emerging technology the implementation of wireless networks and relevant technology was simply an extension of all these health care systems for which nursing staff commonly enjoyed and celebrated the benefit.

> Actually it is the pride to be the leader in the health field; you know, working here as long··· I' ve worked here since late 80s, and we' ve always been the first to try everything, and this is just an extension of that. Nursing administration thinks that computer is telescope, it is a tool to do their jobs; that is the way presented to nursing staffs, and I think that is how it has been accepted through the years··· [Betty, Nursing Analyst]

Moreover, from a competitive standpoint, Urban Healthcare Centre was situated in one of the world's largest medical centres. Numerous surrounding hospitals specialised and competed at many areas as the Centre's. Emerging technology, such as wireless networks in the mid and late 1990s, enabled the Centre to differentiate itself from competitors and helped establish its long-term public reputation such as *Hospitals & Health Networks*'s "100 Most Wired Hospitals," *U.S. News*'s "100 Best Hospitals" (in the U.S.), and *Fortune's* "100 Best Companies to Work for." Particularly in relation to *HHN's (Hospitals and Health Networks)* "Most Wired Hospitals," the winners were being recognised as hospitals that best utilised information technology to address specific areas in the health care service provided (health.usnews.com). The implementation of wireless networks thus helped continue Urban Healthcare Centre's enduring leading status in the field.

> I think Urban Healthcare Centre' s model, as far as they are willing to take a risk to innovate whereas a lot of the institutions will wait until the technologies are stabilized and then they would pursue the implementation of that technology, I think that' s the key differential from us to other institutions··· I think from our perspective, we like to innovate··· Honestly, competition is a big thing in the medical centre. Innovation keeps us ahead of our competitors... [Adam, Project Manager]

As prospective nursing staff continued to recognise Urban Healthcare Centre's innovative technology status in general, and wireless network platform in particular, the Centre built a prosperous relationship with recruiters who could significantly help resolve the nursing shortage issue that commonly existed in the healthcare industry. Internally, wireless networks extended the existing staff members' appreciation of the Centre's IT capability and helped shape their collective pride to be a part of this renowned organisation in the metropolitan area and in the medical field. Externally, wireless networks further legitimised the Centre's innovative status and helped differentiate its competitive advancement from other organisations in the industry as illustrated by Adam's message above.

In reflection, these empirical findings helped answer our research questions, "What are the social values gained from the emergence of wireless networks in organisations?" and "How is organisational legitimacy enhanced through the implementation of wireless networks?" For the former, both organisations' experiences reported here demonstrated various social values that were related to organisational legitimacy. In Metro University, they included potential customers' acknowledgment (i.e. prospective students), internal customers' appreciation (i.e. current staff members or student users), and departmental collaboration (i.e. decentralised IT units across campus). In Urban Healthcare Centre, they were reflected in prospective recruiters' and staff members' recognition (i.e. future nurses), internal customers' satisfaction (i.e. current nurses and doctors), and IT professionals' appreciation for its long-lasting competitive innovation.

For the second research question these organisations' experiences revealed that organisational legitimacy, in relation to wireless networks and IT management, was often enhanced by the recognition and/or appreciation from three major groups of stakeholders: external customers who were normally prospective users, internal customers who were the existing users, and the IT staff members who provided services to the user groups. Prospective customers (i.e. future students in Metro University's case and potential nurses in Urban Healthcare Centre's case) often reflected the general public's perception of an organisation and how it might compete for its future customers. Internal customers' satisfaction or appreciation (i.e. current students in Metro University's situation and nurses in Urban Healthcare Centre's setting) demonstrated how an organisation might retain its existing members and in turn ensure long-term survival opportunity. Finally, IT professionals' appreciation and support of the organisation's IT capability and innovative status often related to how an organisation might retain its technical expertises and subsequently provide better IT services to its prospective and existing organisation members, which in turn helped enhance these two user groups' recognition of the organisation.

Discussion

As these case organisations' experiences demonstrated, wireless networks could provide substantial social values to the organisation. Although these social values might not be tangible or quantifiable in typical economic measures, they often helped an organisation to gain recognition and appreciation from various stakeholders such as prospective customers, internal members, and existing IT professionals. Their heightened realisation of an organisation's IT capability consequently legitimised an organisation's competing position in the field. In the university setting, enhanced organisational legitimacy denoted improved opportunity for recruiting better prospective students externally and satisfying existing students internally. In the healthcare business, it also reflected in greater collaboration with nursing recruiters externally and higher staff satisfaction internally.

For managers, the experiences of Metro University and Urban Healthcare Centre reported here suggested an alternative perspective of IT management in general and wireless network implementation in particular. Traditionally, an organisation often sought economic justification of IT investment. An IT manager might not gain support from chief executive officer (CEO) of chief financial officer (CFO) unless IT projects or investments could provide quantifiable justifications. Those quantifiable justifications would often need to relate to cost reduction or profitability growth. While these quantifiable measures might be essential, managers should not overlook other dimension of IT values that might also help justify project investments. Our empirical findings suggested that certain social values such as prospective customers' recognition, internal members' satisfaction, and existing IT staff members' appreciation might help shape an organisation's long turn image and legitimacy. They might not provide immediate profitability growth but their effects could sustain over time and helped organisations to survive and thrive ultimately.

Particularly for non-profit organisations such as Metro University and Urban Healthcare Centre, profitability growth will become a lesser critical issue than customer services, either internally to its existing staff and IT professionals or externally to its prospective customers. Social values in general, and organisational legitimacy in particular, as illustrated in the case analyses above, were often associated with customer services and in turn helped build an enduring status that might go beyond quantifiable economic measures of IT investment. IT managers in these types of organisations could thus seek an alternative justification (i.e. social justification) to validate their IT investments. More specifically, since wireless networks have rapidly penetrated the society and business world, its social values could help establish an organisation's competing capability and retain prospective and current customers who were increasingly demanding on wireless network services. IT managers could thus evaluate wireless network investment based on social value perspective where the major consideration would be how to enhance organisational legitimacy instead of gaining immediate return on investment.

For researchers, such social value perspectives can help extend the existing body of knowledge in IT and wireless network management. As institutional theorists have argued, an organisation's action might not necessarily result in economics measures. When facing greater technological or environmental uncertainty, organisations' gain in IT investment were often realised in social dimension such as customers services that would in turn sustain an organisation's long term survival. As shown in Metro University's and Urban Healthcare Centre's experiences, organisations in the contemporary wireless world had better seek recognition from various stakeholders because these stakeholders including prospective customers, existing members, and internal IT professionals who were digitally capably and used to ubiquitous information provided to them constantly. Therefore, social justification might be as essential as economic quantification.

Potential avenues for future research might thus include different types of social values that were not manifested in our case studies. For instance, what other social values of wireless networks beyond organisational legitimacy or customer recognition might be realised? From an executive standpoint, questions could also be asked, "How does an IT manager or chief information officer (CIO) validate his/her IT projects in general and wireless network implementation in particular with social values proposed?" In other words, should a CEO or CFO support IT or wireless network investment simply based on social justification? Moreover, while our investigation focused on internal organisation members' realisation of social values resulting from wireless network implementation, in the institutional context, social values might also denote other competing organisations' perspectives in the field. In other words, further investigation could also be carried out in relation to how social values of IT and wireless network investment might shape an organisation's competitive and/or collaborative relationships with other organisations in the field.

Conclusion

Our research investigation set out to help organisations better understand and manage the rapidly developing information and communication technology in general, and wireless networks in particular. Case analyses of Metro University and Urban Healthcare Centre have evidently provided empirical understanding as to how organisations might implement wireless networks to gain social values and in turn enhance its legitimacy in the field. These social values manifested were often associated with prospective customers' recognition, internal members' satisfaction, and existing IT staff's appreciation. Although these social values may not be quantifiable in typical economic measures, their effects could endorse an organisation's ultimate survival in the marketplace. IT managers or researchers should thus pay as much attention to them as to economic values.

However, we do not claim generalisability of our research findings. The case study methodology, particularly an exploratory one, is most suitable for providing practical lessons to which other organisations in similar context might relate. The empirical findings reported here were only based on two case organisations' experiences. They further limit broader implications for other institutional setting or technological environment particularly in for-profit organisations. Furthermore, wireless network technology examined in these case organisations only provided a specific example of a variety of mushrooming ICTs. Future studies could extend research efforts to more cases, from different industries, and/or with other emerging technologies.

In conclusion, our empirical findings suggest that more alternative perspectives of IT management and research are necessary for the business world and the research community. In the contemporary wireless society and business world, the customers that organisations face are digitally capable and wirelessly demanding. For an organisation to survive and sustain in the competitive marketplace, its effort to gain these customers' recognition and appreciation is essential. Their recognition and appreciation might not provide immediate profitability growth or return on investment against which industrial organisations' IT investments were traditionally measured. The implications derived from it, however, could help legitimise an organisation's competing capability and innovative status. Its values might be intangible and non-quantifiable; these social values could lead to an organisation's sustainable success as much as economic benefits could. Organisations situated in today's rapidly changing IT and wireless environment might thus need to reconsider how to better capture various social values derived from IT and wireless network projects and in turn help legitimise its innovative status that might ultimately ensure long-lasting survival.

References

Banker, RD, RJ Kauffman and RC Morey (1990). Measuring gains in operational efficiency from information technology: A Study of the Positran deployment at Hardee's Inc. *Journal of Management Information Systems*, 7(2), 29-54.

Christiaanse, E and J Huigen (1997). Institutional dimensions in information technology implementation in complex network settings. *European Journal of Information Technology*, 6, 77-85.

DiMaggio, PJ and WW Powell (1983). The iron cage revisited: Institutional isomorphism and collective rationality in organizational fields. *American Sociological Review*, 48, 147-163.

Dirsmith, M W., TJ Fogarty and P Gupta (2000). Institutional pressures and symbolic displays in a GAO context. *Organization Studies*, 21(3), 515-537.

Economist (2007a). A world of connections. (cover story), *Economist*, Vol. 383, 3-4: Economist Newspaper Limited.

Economist (2007b). The hidden revolution. (cover story), *Economist*, Vol. 383, 17-18: Economist Newspaper Limited.

Eisenhardt, K M (1989). Building theories from case study research. *Academy of Management Review*, 14(4), 532-550.

Fiser, D. (2004). Wireless technology empowers physicians, *Health Management Technology,* 42.

Glaser, BG and A Strauss (1967). *The discovery of grounded theory: Strategies for qualitative research.* Hawthorne, NY: Aldine Publishing Company.

Hitt, L M, and E Brynjolfsson (1996). Productivity, business profitability, and consumer surplus: Three different measures of information technology value. *MIS Quarterly*, 20(2), 121.

Hoskins, R (2007). Yemen oilfields build WiMAX network, *Broadband Wireless Exchange Magazine.*

Jarvenpaa, S L and DE Leidner (1998). An information company in Mexico: Extending the resource-based view of the firm to a developing country context. *Information Systems Research*, 9(4), 342-360.

King, JL, V Gurbaxani, KL Kraemer, FW McFarlan, KS Raman and CS Yap. (1994). Institutional factors in information technology innovation. *Information Systems Research*, 5(2), 139-169.

Lawrence, TB (1999). Institutional strategy. *Journal of Management*, 25(2), 161-188.

Lofland, J and LH Lofland (1995). *Analyzing social Settings: A guide to qualitative observation and analysis* (3rd ed.). Belmont, CA: Wadsworth Publishing Company.

McCue, A (2007). Thames becomes giant Wi-Fi hotspot.

McKinion, J M, SB Turner, JL Willers, JJ Read, JN Jenkins and J McDade (2004). Wireless technology and satellite internet access for high-speed whole farm connectivity in precision agriculture. *Agricultural Systems*, 81(3), 201-212.

Meyer, JW and B Rowan (1977). Institutionalized organizations: Formal structure as myth and ceremony. *American Journal of Sociology*, 83(2), 340-363.

Miles, MB and AM Huberman (1984). *Qualitative data analysis.* Beverly Hills, CA: SAGE Publications.

Muniwireless.com. 2006. Wifi cities in the US, Vol. 2006.

Oliver, C (1991). Strategic responses to institutional processes. *Academy of Management Review*, 16(1), 145-179.

Patterson, J C (2001). Kids on the run: Mobile technology. *Technology & Learning,* 44-47.

Rajala, JB (2003). Wireless technology in education. *T.H.E Journal*, 31(3), 28.

Rege, O and J Kean (2003). Mobile technology goes to medical schools. *Wireless Systems Design,* 18, 20.

Richards, L (2002). Rigorous, rapid, reliable and qualitative? Computing in qualitative method. *American Journal of Health Behavior*, 26(6), 425-430.

Scott, W R (1987). The adolescence of institutional theory. *Administrative Science Quarterly*, 32, 493-511.

Slack, T and B Hinings (1994). Institutional pressures and isomorphic change: An empirical test. *Organization Studies*, 15(6), 803-827.

Smithson, S and R Hirschheim (1998). Analysing information systems evaluation: Another look at an old problem. *European Journal of Information Systems*, 7, 158-174.

Varshney, U and R Vetter (2000). Emerging mobile and wireless networks. *Communications of the ACM*, 43(6), 73-81.
Yin, R K (1994). *Case study research: Design and methods* (2nd ed.). Thousand Oaks, CA: SAGE Publications.

PART 2

“ R&D ”

- *Technology strategy of R&D Internationalization : An empirical study from a developing country*
- *Russian R&D organizations : case of international technological collaboration*

TECHNOLOGY STRATEGY OF R&D INTERNATIONALIZATION:
AN EMPIRICAL STUDY FROM A DEVELOPING COUNTRY

LOUIS Y. Y. LU

Department of Business Administration, Yuan-Ze University

135, Yuan-Tung Road, Chung-Li, Taoyuan 320, Taiwan

louislu@saturn.yzu.edu.tw

T. M. CHEN

Graduate School of Management, Yuan-Ze University

135, Yuan-Tung Road, Chung-Li, Taoyuan 320, Taiwan

S929608@mail.yzu.edu.tw

World Investment Report 2005 from UNCTAD reveals that "a new trend ... is the emergence and fast growth of foreign R&D activities by TNCs (Transnational Corporations) from developing countries". Numbers of studies have examined several issues of foreign direct R&D investments. Unfortunately, literature in this area has mainly contributed to the knowledge of R&D internationalization of the developed countries. There are only a few studies that have investigated R&D internationalization from the developing countries. Moreover, discussions on paths and technology strategies of overseas R&D establishments over time are still less emphasized.

We examine how Taiwanese firms establish their overseas R&D centers and connect with the global innovation networks over time by conducting both surveys and in-depth interviews. We find four patterns of technology strategy of R&D

internationalization for Taiwanese firms, including the technology-seeking strategy, the technology-exploiting strategy, the technology-harvesting strategy, and the technology-transforming strategy. Most firms present a technology-exploiting strategy, and few belong to a technology-transforming strategy. Moreover, evidence indicates that the most attractive destinations for overseas R&D centers have been shifting from the triad economies into the emerging economies (China, India and Russia).

Keywords: R&D internationalization, Transnational Corporations (TNCs), Developing country, Developed country, Technology strategy

INTRODUCTION

Innovative activity and capabilities are essential for economic growth and development. With the prosperity of globalization, R&D internationalization has a great contribution to enterprises' innovation system. In the 1970s, fewer researchers have been devoted to the field of international R&D investment. Traditionally, R&D activities in TNCs were centralized and concentrated in their home country, mainly because of higher appropriability of R&D efforts (Teece, 1987).Until the early 1990s, the TNCs established an increasing number of R&D laboratories in offshore locations, and researchers also placed greater emphasis on R&D internationalization. A number of studies have examined the issues of foreign direct R&D investment, for instance, the determinants of location (Cantwell and Mudambi, 2000; Jones and Teegen, 2003; Kumar, 2001; Odagiri and Yasuda, 1996; Le Bas and Sierra, 2002), technology transfer (Feinberg and Majumdar, 2001), R&D collaboration (Chen, 1997), the phenomena of R&D clusters (e.g., Silicon Valley and Route 128), the management of multinational R&D (Cheng and Bolon, 1993), and the impact of entry mode and experience-based organizational learning on R&D activities (Belderbos, 2003).

The traditional and main purpose of foreign technological activities is to support foreign production and to service the foreign market (Patel and Vega, 1999). Companies mainly exploit their technological advantage created within their home country. Recently, Internationalizing R&D is often advocated as a strategy for fostering the development of technological capabilities (Penner-Hahn and Shaver, 2005). Niosi (1999) also argues that the role of the internationalization of industrial R&D is changing from technology transfer to the learning organization (Belderbos, 2003). Foreign-based R&D labs have become more involved in technology exploration and advanced development, as opposed to exploitation and adaptation of centrally-advanced, home country-based technologies (Iwasa and Odagiri, 2004; Kuemmerle, 1999; Pearce, 1999). Many researchers also view international R&D as the key to innovativeness and competitiveness in the global knowledge-based and highly-specialized economy (Elder, 2003).

The following two observations underlie the motivations for this study. First, data and research on R&D in developing countries are relatively scattered and few, although international R&D from developing home countries are increasingly valued. Previous studies have indicated that the overseas R&D expenditure of TNCs is highly concentrated in a handful of technologically-advanced regions/countries such as North America, Europe, and Japan (Chen, 2004; Kumar, 2001; von Zedtwitz, 2005), which is nicknamed as the triad. Numerous investigations on R&D internationalization (e.g., Granstrand, 1999; Odagiri and Yasuda, 1996; Patel and Vega, 1999; Ronstadt, 1978) mainly contributed to the knowledge of developed countries. However, the increased activities in international R&D is not limited to just triad firms. Some evidence has indicated that R&D internationalization increasingly involves developing countries, most notably India and China (Chen, 2004; von Zedtwitz, 2005; Wright et al., 2005). Growing numbers of academics and national statistical offices are noticing the rapid increase in overseas R&D activities in the contexts of developing countries. For example, the *World Investment Report 2005* from the United Nations Conference on Trade and Development (UNCTAD, 2005) reveals that "a new trend whereby developing countries are connecting to global knowledge networks is the emergence and fast growth of foreign R&D activities by TNCs from developing countries". A report from the Goldman Sachs Global Research Centers (GS) also predicts that the BRICs (Brazil, Russia, India, and China) economies could become a much larger force in the world economy over the next 50 years (Wilson and Purushothaman, 2003). More research studies to developing countries can enrich the literature of R&D internationalization.

The second observation is that the topic on the evolutionary trajectories of R&D internationalization from a developing country over time is still less stressed in the literature. A substantial amount of research on international R&D, either in the contexts of developing or developed countries, has been analyzed from a static perspective. Von Zedtwitz (2005), for instance, observes the distributions of international R&D from other developing countries in China and proposes two strategies of R&D internationalization from a developing country. One is "Catch up" that describes firms from a developing country conducting R&D in a developed country. The other is "Expansionary" that characterizes a firm in one developing country investing in R&D in another developing country. Nevertheless, von Zedtwitz (2005) ignores how these strategies change over time and how these different paths form and evolve. Although based on a dynamic cross-section model, Cantewll and Piscitello (2000) discuss more on the interrelationship between the diversification, internationalization and accumulation of technological competence in the US. However, more related investigations to the developing countries need to be conducted and reported.

These two observations mentioned above bring us the following research questions. What are the different evolutionary trajectories of firms from a developing country extending their R&D centers to other host countries? How do these paths evolve? What kind of technology strategy do these firms employ in order to upgrade the technological capabilities and connect to global innovation networks? To address these questions, this study surveyed 78 firms that are among the top 1000 manufacturers in Taiwan through either surveys or in-depth interviews. The results reveal four patterns of path and technology strategy – a technology-seeking strategy, a technology-exploiting strategy, a technology-harvesting strategy, and a technology-transforming strategy. Moreover, the most attractive destinations for overseas R&D centers have been shifting from the triad economies into the emerging economies (China, India and Russia).

The remainder of this paper is organized as follows: Section 2 described and summarized prior literature review. Section 3 reports the research methodology and data sample. Section 4 represents the findings of the study and four patterns of R&D internationalization are explained. Finally, the conclusions are disclosed in Section 5.

LITERATURE REVIEW

2.1 MOTIVES FOR ESTABLISHING FOREIGN R&D UNITS

The role traditionally played by R&D laboratories has been that of adapting products for regional or national markets and technology transfer rather than that of undertaking fundamental research (Elder, 2003). This function emphasizes more on market orientation, which is consistent with the definition in Florida's (1997) study and also echoes Vernon's (1966) product life cycle perspective. Elder (2003), and Niosi and Godin (1999) find that for German and Canadian TNCs, technology transfer and adaptation to the local market are still important goals of a foreign R&D establishment.

On the other hand, some research has suggested that internationalization of R&D will in general increasingly become supply-led (Florida, 1997; Granstrand, 1999; Kummerle, 1999) or technological knowledge sourcing (Iwasa and Odagiri, 2004; Patel and Vega, 1999). Expatriate R&D also appeared to be at the root of learning processes through which multinational corporations increased their stock of knowledge in foreign markets (de Mayer, 1993; Niosi, 1999). Kummerle (1999) remarks that the 'capability-exploiting motive for foreign direct investment in R&D' has been the dominant view in international business literature, and characterized the nature of the expatriate technological activities. Granstrand (1999) argues that internationalization of R&D will in general become increasingly supply-led for the reasons of technology diversification and emerging technology market conditions.

Recent literature on the motivations for setting R&D abroad has been revisited, considering the relative competitive advantage and complementary factors. When the host owns the complementary factors that a company needs but the home country does not have, the company will establish R&D units in the host country to acquire the complementary factors, as well as the relative competitive advantage. For instance, Lu and Liu (2004) find that human-capital-augmentation is the major complementary factor for Taiwan's companies in establishing R&D units in China.

2.2 STRATEGY OF OVERSEAS R&D ACTIVITIES

Several R&D strategies have been discussed in the R&D internationalization literature. Tracking back to the missions of R&D activities, Ronstadt (1978: pp. 8-9) investigates 55 R&D units made abroad by seven U.S.-based multinational enterprises and identifies four distinctive kinds of R&D investment abroad: 1) Transfer Technology Units (TTUs) aimed at technical service to the subsidiary; 2) Indigenous Technology Units (ITUs) where new products were developed abroad to service host markets; 3) Global Technology Units (GTUs) where foreign R&D establishments developed products or simultaneous launch in several foreign countries, and 4) Corporate Technology Units (CTUs) where new/basic technology were generated for use by the parent company. Ronstadt finds data showing that most R&D investments were made for manufacturing (TTUs) in order to help transfer the parent's technology and few were for basic research (CTUs).

Bartlett and Ghoshal (1990) proposed four different types of management in international innovation projects: central-for-global (developing new products or processes at home for the global markets), local-for-local (developing products and processes independently in each R&D establishment around the world for use in the local market of the subsidiary), locally-linked (developing a novelty in each location for global exploitation), and globally-linked (developing a novelty through the collaboration of R&D units located in different countries for exploitation in the world market).

Gassmann and von Zedtwitz (1999) developed an evolutionary model of international R&D organization and described five types of international R&D organizations: ethnocentric centralized R&D, geocentric centralized R&D, the R&D hub, polycentric decentralized R&D, and the integrated R&D network. Based on the drivers by access to markets and access to science, von Zedtwitz and Gassmann (2002) proposed four archetypical forms of international R&D organization. National treasure R&D represents domestic research and domestic development. Technology-driven R&D refers to dispersed research and domestic development. Market-driven R&D stands for domestic research and dispersed development. Global R&D symbols dispersed research and dispersed development.

Le Bas and Sierra (2002) investigated 345 multinational firms with the greatest patenting activity in Europe and proposed four types of R&D internationalization: technology-seeking, home-base-exploiting, home-base-augmenting, and market-seeking FDI in R&D. Conducting an empirical study of Japanese R&D investment in the U.S., Iwasa and Odagiri (2002) separated overseas R&D into two types, "research-oriented" and "local-support-oriented." Based on the diversification strategy, Niosi and Godin (1999) differentiated among three types of expatriate laboratories: related diversification, vertically-integrated firms, and global R&D. They find that diversification into related activities is the overseas strategy of Canadian multinational corporations with foreign R&D activities.

Von Zedtwitz (2005) summarized four types of the previous research path in international R&D research. Type 1 concerns the "traditional" R&D internationalization among developed countries, i.e. mostly within the triad of North America, Western Europe, and Japan. Type 2 is the "modern" category of research. The modern form of R&D internationalization became popular in the late 1990s, driven in part by improved economic conditions in South-East Asia, China, and Eastern Europe, in part by strategic considerations of parent companies to set global standards and build global brands, and partly by a growing understanding and financial commitment of TNCs to support local sales with local R&D efforts. Type 3 is "Catch-up" which describes firms from a developing country conducting R&D in a developed country. These firms are naturally attracted to using developed countries as R&D bases, partially in order to acquire local technology and science, to some extent in order to support local product development. The Type 4 strategy of R&D internationalization is "Expansionary," which may be in supporting second-generation technology transfer, or to support other local business activities.

Most of these typologies mentioned above are concluded from data of the triad economies, only a few empirical studies focus on R&D internationalization from the developing countries. Moreover, all research studies on international R&D, either in the context of developing or developed countries, have analyzed this from a static perspective, thus the dynamic and evolutionary process of R&D internationalization over time is still ignored in the literature.

2.3 Overseas R&D activities of Taiwanese manufacturing industry

The R&D expenditures by Taiwanese enterprises are highly concentrated in the manufacturing industry, especially in the electronic and IT industries. Most enterprises still set up their R&D centers in their domestic markets. The R&D expenditures as a percentage of sales to Taiwanese enterprises in the manufacturing sector were low in international standards. Thus, compared to developed countries, Taiwanese R&D activities are less globalized (Lu and Liu, 2004).

The overseas activities of many Taiwanese firms have shifted from exporting, with direct investment in marketing and distribution, to overseas manufacturing, and then, particularly in the high-tech industries, to the start of overseas R&D activity. Generally, most of Taiwan's major sectors are characterized by their vertical disintegration and the pursuit of original equipment manufacturing (OEM) and original development manufacturing (ODM) contracts for brand marketers, without direct access to the final market. The OEM/ODM relationships have also created pressures that forced the Taiwanese firms to extend their production capability rather than to increase their technology innovation capability. Thus, firms based in Taiwan undertook more 'D' than 'R', and that they lack systems integration capabilities (Chen, 2004; Lu and Liu, 2004). However, Taiwanese firms have benefited from OEM/ODM relationships with leading firms and have acquired production engineering capabilities over time.

In 1990s, a few large firms started to internationalize their R&D activities. Some companies, which would go on to build up their own-brand products set up their R&D establishments in the developed countries. Giant and Acer are valid examples, which established their foreign R&D affiliates in the Netherlands and the United States, respectively. Accompanying their industrial upgrades, in the late 1990s, some companies in the semiconductor and IT industries have conducted their overseas R&D labs in the United States. Some companies have established their foreign R&D activities in Japan, Germany, and England. The major missions of these units are to access the local science and to absorb know-how of global value. Since 2000, China has been the most attractive location to most Taiwanese firms. The rapid internationalization in recent years has helped Taiwan obtain a strong position in the list of top 100 developing-economy TNCs. Fifteen companies have entered the list, and all are privately owned and mostly focused in computers and electronics (UNCTAD, 2006).

DATA SAMPLE AND RESEARCH METHODOLOGY

Based on the database of the top 1000 manufacturers in 2005 published by a well-known Taiwanese business magazine, Business Weekly, we screened by means of company publications, news papers, and business magazines to see if these firms have any overseas R&D activities. Then, we contacted firms' spokesperson to confirm the findings. We find that 564 companies out of them are involved in overseas business activities. Among the 564 firms, 242 have extended internationally their R&D activities. Finally, 80 have set up their foreign R&D centers in other developing and/or developed countries. Two out of 80 firms closed in 2006. Thus, 78 firms were chosen as the sample population of this study.

The demographical characteristics of these firms are shown in Table 1.

Table 1. Demographics of sampled firms in 2005.

Item	Scope
Capital	US$12.13 to 5962.5 million
Assets	US$29.875 to 10712.5 million
Sales	US$294.9 to 21046.9 million
Initial Public Offerings	Yes: 62; No: 16

USD: NTD = 1: 32

To learn further detailed information about the locations of R&D centers, motives of overseas R&D activities and technology strategies, we arranged in-depth interviews with R&D managers or top managements.

RESULTS

4.1 FOUR PATTERNS AND TECHNOLOGY STRATEGIES OF OVERSEAS R&D

Extending from the previous studies and considering the trajectory of R&D internationalization from a developing country, we propose four categories of overseas R&D path and technology strategy in the host developing and developed countries. The first pattern is setting R&D centers in the developed countries and staying there consistently to seek the most advanced technology; we call it the technology seeking strategy. The second pattern is establishing R&D centers in other developing countries to exploit the technology they already own, and never try to extend to the developed countries; we label it as the technology exploiting strategy. The third pattern is similar as the first one; however, extending R&D activities to the developing countries later to harvest the knowledge learned from the developed countries, we name it the technology harvesting strategy. The fourth pattern is situating R&D centers in the developing countries firstly, and then extending to the developing countries to gain more advanced knowledge. We title it the technology transforming strategy to indicate these firms try to transform their technology into a higher level. Table 2 demonstrates these four evolutionary paths and technology strategies of R&D internationalization from a developing home country.

Table 2. Patterns of R&D internationalization.

Patterns	Home country	Host country
1.Technology-seeking	Developing	→Developed countries
2.Technology-exploiting	Developing	→Other developing countries
3.Technology-harvesting	Developing	→Developed countries→Other developing countries
4.Technology-transforming	Developing	→Other developing countries→Developed countries

The first pattern refers to those firms from a developing country arranging R&D only in the developed countries, which is consistent with the technology-seeking strategy defined by Le Bas and Sierra (2002). Foreign R&D is used as a vehicle to move closer to the industries of a host country with technological advantages. In comparison with their developed-country counterparts, TNCs from developing economies have generally been technological followers. To improve their competitiveness, especially in terms of seeking to compete in global markets, firms based in emerging economies may be forced to enter developed countries by acquiring new resources and capabilities (Cantwell, 1992; von Zedtwitz, 2005), or by exploration to build up potential absorptive capacity (Wright et al., 2005). Firms from the developing home countries have conducted innovative R&D abroad in the developed countries in order to tap other centers of innovation and overcome the constraints of their domestic economy (Criscuolo, et al., 2005). However, their primary stimulus of this R&D internationalization is to "catch-up" (von Zedtwitz, 2005) with developed countries, such as R&D sites of Taiwanese semiconductor companies in the US. Typically, enterprises under this pattern tend to be more technological innovation orientated.

Pattern 2 represents a technology-exploiting strategy – firms from a developing country conduct R&D in other developing countries. The overseas R&D centers under this pattern exhibit the characteristics of the technology transfer units (TTUs) of Ronstadt's typology. These centers may support second-generation technology transfer (Chen, 2004; Ronstadt, 1978; von Zedtwitz, 2005) or in acquiring human capital (Lu and Liu, 2004). In this type, companies are active in foreign locations in technical fields where they are relatively strong at home. Moreover, von Zedtwitz (2005) argues that firms of developing countries will internationalize R&D into other developing countries opportunistically. As a consequence, they may reap long-term first mover advantages in less privileged regions of the world.

Pattern 3 represents a technology-harvesting strategy, i.e. firms conduct R&D in developed countries first and then establish R&D centers in the developing country later– also namely, a technology seeking-exploiting strategy. Companies in this pattern develop the R&D network's capabilities to understand and conduct cutting-edge technology development by absorbing know-how from developed countries, and then passing on the absorbed technologies and technical know-how and refined for use in other developing countries.

Pattern 4, the technology-transforming strategy, describes firms arranging R&D in other developing countries early and then entering developed countries later. Companies under this pattern are mostly in a mature industry and would like to upgrade their technology level to enter a new industry. This is similar to intersectional upgrading, which is applying the competence obtained in a particular function to move into a new sector. In this pattern, technological capabilities in the developed host location become increasingly important over time. Basic research is the most important mission of the units, and thus the cost of the host location is not a major concern in the location decision.

After in-depth interviews, we find that 75 Taiwanese firms can be grouped into four patterns of R&D internationalization, while three companies present a random and complicated type. We hence only examine the 75 firms and differentiate among the four patterns of overseas R&D centers. Table 3 and Figure 1 present the details of four paths and technology strategies of R&D internationalization from 1990 to 2005.

Table 3. Patterns of R&D internationalization of 75 firms.

Patterns	No. of firms	Percentage (%)	Capital (US$ millions)
Pattern 1: Technology-seeking strategy (Taiwan → Developed countries)	20	26.67	12.13 ~ 1093.75
Pattern 2: Technology-exploiting strategy (Taiwan → Other developing countries)	32	42.67	30.19 ~ 5962.50
Pattern 3: Technology-harvesting strategy (Taiwan → Developed countries → Developing countries)	21	28.00	17.40 ~ 1559.34
Pattern 4: Technology-transforming strategy (Taiwan → Developing countries → Developed countries)	2	2.67	42.13 ~ 1562.50
Total	75	100	

According to Table 3, most of the firms belong to a technology-exploiting strategy, occupying 43%. Only two percent of firms are under a technology-transforming strategy. We find that since the early 1990s, twenty out of 75 firms under a technology-seeking strategy have located their foreign R&D centers only in developed countries. Twenty-one firms, classified under a technology-harvesting strategy, established their international R&D centers in developed countries in the early 1990s and then shifted their next R&D centers in the developing countries. Thirty-two firms belong to a technology-exploiting strategy, which have started to set up R&D units in other developing countries since the late 1990s. Only two cases adopted a technology-transforming strategy, which built up overseas R&D centers in other developing countries in the late 1990s and recently established R&D sites in developed countries. The results react to the facts that internationalization of R&D has spread to developing countries, and firms from the developing country are themselves investing in R&D either in developing countries or developed countries.

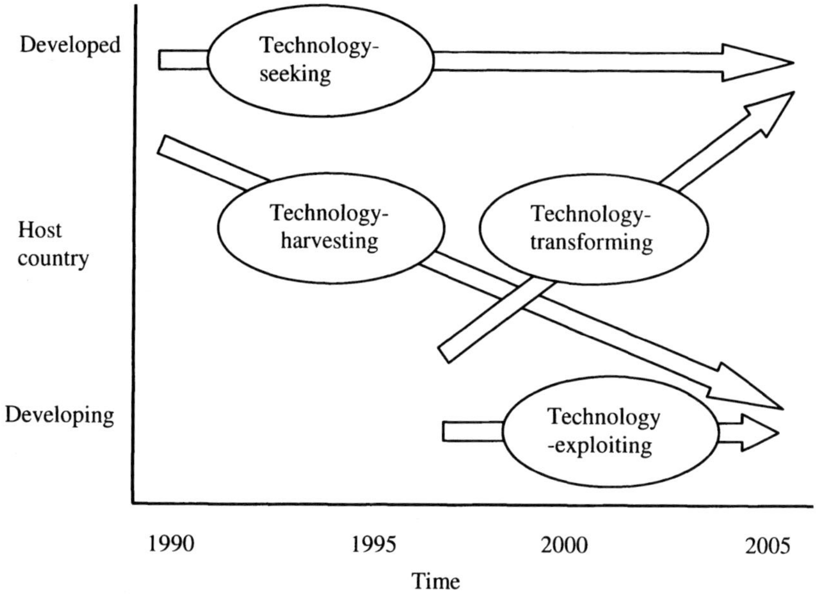

Figure 1. Strategy of R&D internationalization from 1990-2005

Table 4. Distribution of 75 firms: pattern and principal product group.

Product group	Pattern 1	Pattern 2	Pattern 3	Pattern 4	Total
IT	4	24	12	1	41
Semiconductor	6	1	4	-	11
Electric equipment	2	2	3	-	7
Photoelectric	2	1	1	1	5
Pharmaceuticals	3	-	-	-	3
Automotive and Transportation	-	2	1	-	3
Chemicals	1	1	-	-	2
Rubber	1	-	-	-	1
Furniture	-	1	-	-	1
Food	1	-	-	-	1
Total	20	32	21	2	75

In Table 4, we list the number of 75 firms according to the pattern and principal product group. In terms of the industrial distributions, we find over half (55%) of firms from the IT industry, followed by 15% from the semiconductor industry, 8% from the electric equipment industry, and 7% from the photoelectric industry. The findings imply that R&D activities in the IT industry are more internationalized than those in the other industry. In the IT industry, 24(54%) out of 41 firms are under pattern 2, which takes a technology-exploiting strategy, followed by pattern 3. In the semiconductor industry, 6(55%) out of 11 firms take a technology-seeking strategy, and 4 firms are under a technology-harvesting strategy. Under pattern 1, the results show 30% of firms from the semiconductor industry, followed by those from the IT and pharmaceutical industries. We discover that 75% and 55% of firms under pattern 2 and pattern 3 are from the IT industry, respectively.

Table 5. Paths of R&D centers in the host countries.

Patterns	Number of firms
Pattern 1	
United States	15
United States→Germany	2
Japan	2
Denmark	1
Pattern 2	
China (including Hong Kong)	27
China→Malaysia	2
China→Vietnam	1
China→Thailand	1
Vietnam	1
Pattern 3	
United States→China	7
United States→China→India	3
United States→China→Malaysia	2
United States→Singapore	1
Canada→China	1
Japan→China	3

Netherlands→United States→China	1
Germany→Russia	1
Germany→Japan→United States→Russia	1
England→India	1
Pattern 4	
China→Japan/Netherlands	1
Malaysia→China→Japan	1
Total	75

4.1 PATTERN 1: TECHNOLOGY-SEEKING STRATEGY

Table 5 reveals that fifteen out of 20 firms have set up their R&D centers in the United States. Two companies have first established their R&D centers in the United States and later in Germany. Two firms and one firm have built up their international R&D centers in Japan and Denmark, respectively.

Under this pattern, we find that the most important location of R&D centers in developed countries is the U.S. followed by Germany and Japan. The U.S. attracts 17 out of 20 firms to set up their foreign R&D centers, especially those in the IT, semiconductor and pharmaceutical industries. The results reveal the close technological relationships between Taiwan and the United States. In-depth interviews revealed that these firms will continually internationalize R&D into the triad economies, because of the shortage of domestic technologies and the leading position of the advanced host countries in some technological fields.

4.2 PATTERN 2: TECHNOLOGY-EXPLOITING STRATEGY

Table 5 shows that 27 companies out of the 32 firms set up their R&D centers directly in China. Most of them are categorized in the IT industry. Four companies in the IT and automotive industries established their overseas R&D centers firstly in China and then subsequently in South-East Asia (i.e. Malaysia, Vietnam, and Thailand). Only one automotive company situated its R&D center in Vietnam.

Since the mid-1990s, electrical and electronic equipment manufacturers in Taiwan have improved their competitiveness by investing in China in association with production, followed by investment in R&D. We observed that overall 31 firms have set up their international R&D centers in China. Among them, we find 17 companies with single-site R&D centers and 14 with multi-site R&D centers in China. These R&D centers – mostly established in Beijing, Shanghai, and Tianjin, as well as Guangdong and Jiangsu provinces – permit multinationals closer access to their large Chinese customer base.

Although pattern 2 has an emphasis on the technology exploiting tendencies of firms from a developing country conducting R&D centers in other developing countries, Lu and Liu (2004) argue that local effective human capital provided by China is the major complementary factor for their R&D extensions for two reasons. The first is that most Taiwanese companies rely more on OEM and/or ODM contracts with major global companies. Only a few companies promote their brand products in China. Even though China's market offers big potential, it has not been a major market emphasis to most Taiwanese firms. Thus, immediate local support is the main factor, but not product and service localization. The second is that China offers Taiwan well-educated local engineers, which are cost effective and share geographical and linguistic proximities.

We agree overall that the most important incentive to Taiwanese firms conducting R&D in China is the human-capital-augmentation, while that in South-East Asia (Malaysia, Thailand, and Vietnam) is the tendency for a technology-exploiting strategy.

4.3 Pattern 3: technology-harvesting strategy

Table 5 also demonstrates the distributions of pattern 3 of foreign R&D centers. Twelve companies conducting R&D in the triad economies (Europe, North America, and Japan) in the beginning have built up their next R&D centers in China. Three firms situated their R&D centers in the U.S., followed by China and India. Two firms set up their R&D in the U.S., China, and then later in Malaysia. One conducted an R&D center in the U.S. then invested in Singapore. Overall, 20 firms have set up their international R&D in Russia, India, and China and the result reveals that BRICs (except Brazil) are the most attractive emerging countries for Taiwanese firms to internationalize their R&D centers. In-depth interviews disclose that Brazil is too far away for Taiwanese firms and lacks incentives to set up their R&D center.

Two of the three companies establishing R&D centers in the triad economies first set up their follow-up R&D centers in Russia and another one in India. Because of the abundant highly trained cheap manpower and the existence of a few internationally known public-funded centers of excellence, R&D managers in semiconductor companies mention that India has attracted firms to set up R&D and software development centers. However, we find that two companies closed their R&D centers in the U.S. Top management of these companies provided two explanations for this phenomenon. The first is the high operating costs of the R&D centers in the U.S. The performance of international R&D in developed countries cannot be realized in the near term. The second is the rapid emergence of BRICs which have improved their technological capabilities gradually and provided well-educated human capital.

We also find that most of these companies in the IT industry under this pattern seem to be the first movers to internationalize their R&D into the triad economies and are likely to build up their own brand products. In the host developed countries, these well-known developing-country firms moved up along the value chain to create their brands, thus becoming similar to major developed-country TNCs. Recently, these firms have committed to R&D inputs in other developing countries. For instance, one company in the transportation industry set up their R&D centers in the Netherlands in 1986 and then in the US. The purpose of the R&D units is to learn the pioneering technology from these countries and then to pass the learnt knowledge and technical know-how to its parent company. Recently, this company creates a new R&D center in China to support local production.

4.4 PATTERN 4: TECHNOLOGY-TRANSFORMING STRATEGY

In Table 5, there are only two cases of the transformation strategy. These two firms are in a mature product line of the electronic industry and enter another new industry to upgrade their technology capabilities through mergers and acquisitions (M&As) One of the firms is originally a mobile display company which has set up their foreign R&D centers in China and has acquired its other overseas R&D related to the photoelectric industry in Japan and Netherlands. The other one, which is primary a cellular phone manufacturer, conducts its overseas R&D in Malaysia and China first and then obtains an R&D center related to the telecommunication industry in Japan. A top manager of these companies indicates that the exploitation of synergy effects is a principal motivation for the M&A activity. The M&A route is more attractive where speed in accessing the technology or innovative strengths in a host developed economy are primary benefits. For instance, one company in the mobile display industry mentioned above completed a merger in 2005. Two partners are involved: one partner from the Netherlands has been in the display business for more than 45 years and the other partner from Taiwan has successfully pioneered LTPS (Low Temperature Poly Silicon) technology in the flat panel industry. Through the combined strengths and synergies, the company has evolved into an industry leader in the global mobile display industry. This finding also reveals that few Taiwanese firms are able to enter a new technological field through M&A.

CONCLUSION

The purpose of this study is to throw new empirical light on the internationalization of R&D activities from a developing country by Taiwanese firms. The most important potential home-country gain from outward R&D investments is the improved competitiveness and performance of the firms and industries involved. Although the overseas R&D activity of Taiwanese manufacturing firms may still be in a rather premature stage compared with that of the triad economies, the survey data reveal the fact that foreign R&D activities have been increasing in recent years. We categorize herein four patterns of technology strategy of R&D internationalization from a developing country, including the technology-seeking strategy, technology-exploiting strategy, technology-harvesting strategy and technology-transforming strategy. We also find that thirty-two firms belong to a technology-exploiting strategy, which have only entered other developing countries, especially China, and only two firms correspond to a technology-transforming strategy. The results show that in the early 1990s, the triad economies were the most important R&D sites for Taiwanese firms. In the late 1990s, the emerging economies (e.g. China, India, and Russia) gradually become the major destinations with Taiwanese overseas R&D affiliates.

To most Taiwanese firms conducting a technology-seeking strategy, we discover the fact that the most important host country is the U.S., followed by Japan and Germany. The technological capabilities in the triad economies lead those in Taiwan. Thus, Taiwanese firms with innovation orientation tend to choose the triad economies as their first destinations. This finding echoes the observations on clustering which indicate that significant economies of agglomeration exist in the geographical location of national scientific capacity (Audretsch and Feldman, 1996; Baptista and Swann, 1998, Reddy, 2005).

In the other host developing countries, three of the BRICs (except Brazil) are the most attractive emerging countries for Taiwanese firms to internalize their R&D centers. The rapid emergence of BRICs is because these emerging countries have improved their technological capabilities gradually and provided well-educated human capital. Among them, China is the major destination for most Taiwanese firms in the IT industry. Overseas R&D centers in China are concentrated in large cities with skilled human capital, particularly in Beijing and Shanghai. Some managers indicate that strategic foreign R&D in China does exist and is not only involved in short-term production development, but also in long-term original research. The mission of these R&D laboratories in China is to become an international R&D center, rather than a support laboratory serving the local market. Thus, these R&D centers value not only the Chinese market, but also available talents and technological capacities.

Evidence suggests that companies from a developing country, which would like to build up their own brand products, can internationalize their R&D into the triad economies first to enhance their innovation capacities and then commit R&D inputs in other developing countries to serve the local market. Through creating the R&D network's capabilities to understand and conduct original technology development by learning know-how from developed countries, these firms promote along the value chain to construct their brands, thus becoming similar to developed-country TNCs. In the past, international M&A may exist more when large TNCs from a developed country enter a host developing country and obtained a local company, while the data show that two companies acquire their foreign R&D centers of the developed countries via the M&A to upgrade their technological level and transform into another new industry. Thus, M&A is a possible entry mode for firms from a developing country to speed up access to the technology or innovative strengths in a host economy,

There are two opposing views regarding the impact of TNC's R&D on the host countries. One view considers inward R&D-related FDI to be beneficial to economic growth, but the counter view argues that R&D activities by foreign firms tend to tap into unique local R&D resources with little or no benefit to the host country (Reddy, 2005). However, countries have individual innovation capacity that influences their R&D activities both at home and abroad (Nelson, 1993). To policy makers, it is important to understand the nature of the country-specific factors that have an influence in creating national technological advantages, including the competitive climate, the financial system, and education, training and basic research institutions. They should develop realistic attitudes as to what kind of R&D foreign direct investment they can attract to their nations and regions.

In summary, this study enriches the understanding of R&D internationalization and makes a contribution to the area of technology strategy of R&D internationalization from a developing country over time. Four patterns of technology strategy – 'technology-seeking strategy', 'technology-exploiting strategy', 'technology-harvesting strategy', and 'technology-transforming strategy' – typify R&D internationalization from a developing country. The most attractive destinations for overseas R&D centers have been shifting from the triad economies into the emerging economies (e.g. China, India, and Russia).

Acknowledgements

The authors would like to thank the National Science Council, Taiwan, for funding the project coded by NSC 95-2416H-155-008.

REFERENCES

Audretsch, D. B. and M. P. Feldman (1996). Innovative clusters and the industry life-cycle, *The Review of Industrial Organization*, 11, 253-273.

Baptista, R. and G. M. P. Swann (1998). Do firms in clusters innovate more? *Research Policy*, 27(6), 525 – 540.

Bartlett, C. A. and S. Ghoshal (1990). Managing innovation in the transnational corporation. In *Managing the Global Firm*. C.A. Bartlett, Y. Doz and G. Hedlund (eds.) London: Routledge.

Belderbos, R. (2003). Entry mode, organizational learning, and R&D in foreign affiliates: evidence from Japanese firms, *Strategic Management Journal*, vol. 24, 235-259.

Cantwell, J. (1992). The theory of technological competence and its application to international production. In *Foreign Investment, Technology and Economic Growth,* D. McFeteridge (ed.), pp. 33-67. University of Calgary Press.

Cantwell, J. and R. Mudambi (2000). The location of MNE R&D activity: The role of investment incentives, *Management International Review*, 40(1), 127-148.

Cantwell, J. and L. Piscitello (2000). Accumulating technological competence: It's changing impact on corporate diversification and internationalization, *Industrial Corporation Change*, 9, 21-51.

Chen, S. H. (1997). Decision-making in research and development collaboration, *Research Policy*, 26(1), 121-135.

Chen, S. H. (2004). Taiwanese IT firms' offshore R&D in China and the connection with the global innovation network, *Research Policy*, 33(2), 337-349.

Cheng, J. and D. S. Bolon (1993). The management of multinational R&D: A neglected topic in international business research. *Journal of International Business Studies*, 24(1), 1-18.

Criscuolo, P., R. Narula, and B. Verspagen (2005). Role of home and host country innovation systems in R&D internationalization: a patent citation analysis, *Economic of Innovation and New Technology*, 14(5), 417-433.

De Meyer, A. (1993). Internationalization of R&D improves a firm's technological learning, *Research-Technology Management*, 36(4), 42-49.

Edler, J. (2003). Germany and the internationalization of industry R&D: New trends and old patterns. In *Multinational Enterprises, Innovation Strategies and Systems of Innovation*. J. Cantwell and J. Molero (eds.) Cheltenham: Edward Elgar Publishing Limited. pp. 105-128.

Feinberg, S. E. and S. K. Majumdar (2001). Technology spillovers from foreign direct investment in the Indian Pharmaceutical industry, *Journal of International Business Studies*, 32(3), 421-437.

Florida, R (1997). The globalization of R&D: Results of a survey of foreign-affiliated R&D laboratories in the USA, *Research Policy*, 26(1), 85-103.

Gassman, O. and M. von Zedwtitz (1999). New concepts and trends in International R&D organization, *Research Policy*, 28(2), 231-250.

Granstrand, O (1999). Internationalization of corporate R&D: A study of Japanese and Swedish corporations, *Research Policy*, 28(2), 275-302.

Iwasa, T. and H. Odagiri (2004). Overseas R&D, knowledge sourcing, and patenting: An empirical study of Japanese R&D investment in the US, *Research Policy*, 33(5), 807-828.

Jones, G. K. and H. J. Teegen (2003). Factors affecting foreign R&D location decisions: management and host policy implications, *International Journal of Technology Management*, 25, 791-813.

Kuemmerle, W (1999). The drivers of foreign direct investment into research and development: An empirical investigation, *Journal of International Business Studies*, 30(1), 1-24.

Kumar, N (2001). Determinants of location of overseas R&D activity of multinational enterprises: The case of US and Japanese corporations, *Research Policy*, 30(1), 159-174.

Le Bas, C. L. and C. Sierra (2002). Location versus home country advantages in R&D activities: Some further results on multinationals' locational strategies. *Research Policy*, 31(4), 589-609.

Lu, L. Y. Y. and J. S. Liu (2004). R&D in China: An empirical study of Taiwanese IT companies, *R&D Management*, 34(4), 453-465.

Nelson, R. (1993). *National Systems of Innovation: A Comparative Study*. Oxford: Oxford University Press.

Niosi, J (1999). The internationalization of industrial R&D from technology transfer to the learning organization, *Research Policy*, 28(2), 107-117.

Niosi, J. and B. Godin (1999). Canadian R&D abroad management practices, *Research Policy*, 28(2), 215-230.

Odagiri, H. and H. Yasuda (1996). The determinants of overseas R&D by Japanese firms: An empirical study at the industry and company levels, *Research Policy*, 25(7), 1059-1079.

Patel, P. and M. Vega (1999). Patterns of internationalization of corporate technology: Location vs. home country advantage, *Research Policy*, 28(2), pp. 145-155.

Pearce, R. D. (1999). Decentralized R&D and strategic competitiveness: Globalized approaches to generation and use of technology in multinational enterprises (MNCs), *Research Policy*, 28(2), 157-178.

Penner-Hahn, J. and J. M. Shaver (2005). Does international research and development increase patent output? An analysis of Japanese pharmaceutical firms, *Strategic Management Journal*, 26(2), 121-140.

Reddy, P (2005). R&D-related FDI in developing countries: implications for host countries, *Proceedings of the Expert Meeting*, pp. 97-105.

Ronstadt, R.C. (1987). International R&D: The establishment and evolution of research and development abroad by seven U.S. multinationals, *Journal of International Business Studies*, 9(1), 7-24.

Teece, D. J. (1987). Capturing value from technology innovation: Integration strategic partnering and licensing decision. In *The Competitive Challenge,* D.J. Teece (ed.) New York: Harper and Row.

UNCTAD (2005). *Transnational Corporation and the Internationalization of R&D.* New York and Geneva: United Nations.

UNCTAD (2006). *FDI from Developing and Transition Economies: Implications for Development*, New York and Geneva: United Nations.

Vernon, R. (1966). International investment and international trade in the product cycle, *The Quarterly Journal of Economics*, 80(2), 190-207.

von Zedtwitz, M. (2005). International R&D strategies in companies from developing countries — The case of China, UNCTAD.

von Zedtwitz, M. and O. Gassmann (2002). Market versus technology driven in R&D internationalization: Four different patterns of managing research and development, *Research Policy*, 31(4), 569-588.

Wilson, D. and R. Purushothaman (2003). *Dreaming with BRICs: The Path to 2050.* Goldman Sachs Global Research Centers.

Wright, M., I. Filatotchev, R. E. Hoskisson and M. W. Peng (2005). Strategy research in emerging economies: Challenging the conventional wisdom, *Journal of Management Study*, 42(1), 1-33.

RUSSIAN R&D ORGANISATIONS: CASES OF INTERNATIONAL TECHNOLOGY COLLABORATION

ANNA TRIFILOVA

Nizhny Novgorod Architecture and Civil Engineering State University,

Management and Marketing Department,

E-mail: anna-tri@yandex.ru

Using a case-study approach, this investigation examines international technology partnerships taking Russian R&D organisations for a research sample. The paper has been inspired by the introduction of the liberalisation policy in the Russian Science and Technology (S&T) sector and proliferation of the globalisation approaches in technology management. The central idea of the study is to research international experience gained by Russian R&D organisations after the collapse of the Soviet Union and the emergence of the market economy in Russia. The main research goal is to understand perspectives of Russian R&D organisations in establishing partnerships with international companies. The scope of the paper is to gain insight into the nature of organisational inertia hindering Russian R&D establishments from expanding globally through technology partnerships. The paper reports on a comparison of two innovation-oriented companies having experience in customer-supplier and joint venture organisational modes of international technology partnerships. The comparison is based on semi-structured interviews focusing on (1) source, (2) recipient, (3) technology, (4) management, and (5) country-specific attributes selected from the literature. The paper combines a variety of empirical challenges Russian R&D organisations experience going globally. The results reveal that Russian R&D organisations encounter three groups of issues that could be grouped as: (i) neutral, (ii) contributing, or (iii) hindering the establishment of international technology collaboration. The study contributes to the field in technology management through the exploration of inertia Russian science-intensive firms encounter entering global R&D market. Delineation of organisational inertia contingent on modes of international technology partnership is for policy-makers to call for different management attention in the initiation of cooperation projects with R&D organisations from emerging market economies.

Keywords*: technology partnerships; cooperative modes; inertia; R&D organisations; Russia*

Introduction

Technology collaboration refers to how organisations co-operatively create and commercialise new technologies. A number of constituent changes in the landscape of technology management have combined recently to underpin investigation of international co-operation with Russian R&D organisations. These underpinnings are emergence of the market economy in the ex-USSR (Union of Soviet Socialist Republics), liberalisation policy in Russian S&T, internationalisation of R&D, globalisation of technology, and open innovation paradigm.

Prior to the market reforms of the 1990s, Russian S&T sector was centrally directed and totally financed by the Soviet Government. R&D was supplied to industrial enterprises as a free good of the centrally-planned economy and all inventions were state property. In the USSR intellectual state property was freely available to use for anyone without licences or royalty payments, provided that such usage was deemed to be in the interests of the country. In the Soviet period, an inventor received public recognition in the form of an Authors' Certificate. Under no circumstances, the Authors' Certificate granted the inventor an exclusive right for patent protection (Watkins, 2003). The transition from the planned to market economy put a stop to the command system of accomplishing R&D. Ex-USSR R&D organisations gained liberalisation for the whole innovation process starting from producing a novel idea to taking it to implementation and making a profit from it (S&T Overview, 2006). The initial structural shifts that faced S&T in the journey to the market economy were a tremendous decrease in federal budget expenditure on R&D and a complete disintegration of hierarchical administrative system.

At the beginning of the 1990s, the shift from the state controlled R&D system of the Soviet Union to the market-based economy of the Russian Federation seemed to have generated enormous business opportunities for Russian R&D organisations. Having lost the Soviet Government as their only R&D funder and contractor, ex-USSR R&D organisations started searching for new partnership perspectives. Comparing with the time of the iron curtain, the prospects of Russian R&D organisations for establishing technology collaboration and commercialising innovations seemed to be boundless. The most lucrative markets for Russian science-intensive enterprises and research institutes considered to be "outside of the country, either in other emerging economies or occasionally, in Western Europe or the USA" (Watkins, 2003), though, only a few of the Russian R&D organisations managed to find partners for technology cooperation overseas (Powell, 2005; Trifilova 2007a). Empirical investigations of Russian R&D sector lead to the observation that "Russian innovative companies are seldom looking for collaborative opportunities externally...explaining it by a lack of market knowledge and management skills necessary to develop internationally" (Trifilova *et al.*, 2007).

The paper at hand focuses on international technology collaboration of ex-USSR R&D organisations studying experience gained by Russian R&D organisations in the global R&D market after the collapse of the Soviet Union and the shift to market economy in Russia. Investigation of international (rather than national) co-operation strategies of Russian R&D organisations aims to highlight their collaboration perspectives on the global market and goals to understand inertia hindering innovation-oriented establishments from expanding externally in the time of liberalisation policy in the Russian Federation. To achieve this objective the paper is organised as follows. It begins with a brief discussion of the literature review on international technology co-operation and a sketchy description of the conceptual foundations of Russian S&T sector. The paper subsequently goes on to define research methods and present main findings. The final section draws conclusions on major results of the study. The analysis here is derived from an ongoing research of international experience of Russian R&D organisations (Trifilova 2006; 2007a; 2007b; Trifilova *et al.*, 2007) and further contributes to the knowledge of technology management in emerging market economies.

Literature Review and Empirical Background

The notion of technology co-operation is widely described as an interaction between organizations that is largely beneficial to all those participating in it. A decision to co-operate is generally considered as a strategic intent, which aims at accomplishing more with fewer resources and at a faster pace (Todeva and Knoke, 2005).

A. Chiesa et al. *(2000) provide a comprehensive summary of inter-organizational relationship modes of technology collaboration describing the roles of actors in it. The mostly used approaches in classifying technological relations are equity involvement and time horizons. When a company agrees with others jointly to carry out R&D on a definite technology with no equity involvement, it is considered as joint R&D (Chiesa* et al.*, 2000). When a company agrees to fund the cost of R&D at a research institute, university or small innovative firm for a definite technology, it is treated as R&D contract. When a company shares technology resources with other companies in order to achieve a common objective of technological innovation without equity involvement, the notion of alliance is used. In the study to follow customer-supplier (i.e. R&D contract) and joint-venture modes are examined.*

Motivations for internationalising R&D include a gateway to scientific talent and international division of labour (Nobel and Birkinshaw, 1998); modifications of companies' existing products to better fit local market needs as well as collection of direct feedback from customers and technical support for local production (Gunasekaran, 1998); knowledge-sharing and higher innovation performance (Spencer, 2003); quick entry into new technological fields (Belderbos, 2003); keeping pace with technology development in the host country and co-operate with foreign R&D labs (Dalton *et al.*, 1999). The main objective on investigating partnership with Russian R&D organisations, belonging to emerging market economies, for the study to follow is placed not on the motives *per se* rather on the variables contributing to the establishment of international technology collaborations and hindering the development of co-operation as in Farr and Fisher (1992), Hakanson (1993), Teichert (1993), Tyler and Steensma (1995), Bruce *et al.* (1995), Kim and Lee (2003), Narula (2004, 2005).

Russian R&D organisations are little-studied in the context of integration into international technology cooperation milieu mainly for historical reasons. Emphasising R&D efforts on the basic research with space and army applications, the strength of the S&T sector in Soviet period was widely considered of great importance among political objectives and considerations of international prestige (S&T Overview, 2006). It was the time of excessive secrecy, location of many R&D assets in closed atomic cities, scarcity of communication between R&D establishments on all levels, along with no financial limitations in terms of budgetary support (Watkins, 2003). Soviet system created a unique R&D base which was military structured and highly educated, geographically isolated and extremely large, functionally separated (Radovilsky, 1994) and required a thorough reorganisation when adjusting to the market economy. Paying little attention to economic return, commercial orientation, needs of the consumer market or industrial innovations, Soviet S&T ill-suited liberalisation policy commenced in the country at the beginning of 1990s (Gokhberg and Shulanova, 2004).

Since the beginning of the economic reforms Russian S&T sector and ex-USSR R&D organisations have been trying to narrow the gap between the legacies of the Soviet command system and the market policy. While the government is transforming S&T sector and adjusting it to the knowledge economy, R&D organisations making own efforts to use their research capabilities (Gokhberg *et al.*, 2001) and technological potential to overcome inherited inefficiencies. Adjusting to the demands of the market economy, Russian science-intensive enterprises and research institutes require production and/or marketing partners that can help develop and expand consumer-demanded rather than military-oriented innovations. After two decades of liberalisation policy in post-Soviet S&T sector, this paper investigates experience of Russian R&D organisations in establishing external partnerships. Aiming to explore hindrances Russian R&D organisations encounter internationally after the introduction of liberalisation policy in the Russian S&T sector, the study poses the following research question:

Understanding the inertia hindering Russian R&D organisations from expanding globally as an emerging technology supplier or a strategic R&D partner.

Methodology

Investigation was accomplished using a case-study method. The choice of the qualitative analysis considered advantageous for the discovery of how variables contingent on the establishment of technology partnerships are adjusted for Russian R&D organisations (Yin, 2003). The primary research instrument was an interview protocol comprising general and specific questions. First the interview was focused on the background information about the interviewing company. Then it moved to a retrospective analysis of the experience in international partnership using variables given in table 1.

Table 1. Variables contributing to technology collaboration

Group	*Factor*	*Literature*
Source	Cooperative history	Bruce et al., (1995); Hakanson (1993); Dodgson (1993); Tyler&Steensma (1995)
	Expertise level	Gunasekaran (1998); Matthews (2006); Trifilova (2007*b*); Watkins (2003); Powell (2005)
	IP rights	Harris *et al.* (1996); Brooking (1996); Teece (2000); Watkins (2003); MacDougall&Hurst (2005)
Recipient	Compatible goals	Teichert (1993); Hausler et al., (1994); Saxton (1997); Mohr&Spekman (1994); Stamm (2004)
	Contribution	Dodgson (1993); Tyler&Steensma (1995; 1998); Bruce *et al.*, (1995); Alcorta *et al.*, (1998)
	Patterns of payoffs	Parkhe (1993); Kim&Lee (2003); Rothaermel&Deeds (2004); Chiesa *et al.* (2005)
Technology	Further development	Tyler&Steensma (1995); Rothaermel&Deeds (2004); Piachaud (2005)
	Expected cost	Tyler&Steensma (1995); Gunasekaran (1998); Rothaermel&Deeds (2004); Chiesa *et al.* (2005);
	Life cycle	Hagedoorn (1993); Ming&Xing (1999); Lin *et al.* (2002) Poulymenakou&Prasopoulou (2004)
Management	Coordination	Souder&Nassar (1990); Alcorta *et al.*, (1998); Parker (2000); Kleinsmann&Valkenburg (2005)
	Mutual trust	Dodgson (1993); Humphrey&Schmitz (1998); Das &Bing-Sheng (1998) Carneiro (2006); Swan&Newell (1994); Bennett *et al.* (1997); Narula (2005); Todeva&Knoke (2005)
	Mentality	

Country-specifics	Communication	Sakakibara (1993); Bruce *et al.*, (1995); Ragatz *et al.*, (1997); Parker (2000); Narula (2003, 2005)
	Cultural differences	Alcorta *et al.*, (1998); More&McGrath (1999); Gunasekaran (1998); Piachaud (2005)
	Marketing experience	Farhang (1997); Czuchry&Yasin (1999); Zain&Yasin (1997); Narula (2003); Trifilova *et al.* (2007)

Table 1 summarizes variables contributing to technology collaboration grouped into (1) source, (2) recipient, (3) technology, (4) management, and (5) country-specific attributes. Different challenges emphasize various factors that if in the positive – are associated with the contribution to the success of the co-operative project, and if in the negative – are hindering the establishment of technological partnership. The list of factors is not claimed to be universal and complete). It is derived from the literature review and preliminary study and is used as a set of research variables for an in-depth case-study analysis in the research to follow. To avoid biasing the notion 'inertia' was never used by the interviewer.

The research sample was delimited to two Russian R&D organisations having experience in customer-supplier and joint venture co-operative modes. Comparison of the studied cases reveals challenges relevant to different patterns of technology partnerships. Unlike survey research, which relies on quantitative generalisation, the emphasis of this study is more on analytical generalisation. The scope of the paper is rather to gain insights into the process of establishing technology partnerships with Russian R&D organisations than in producing statistically important results.

Research Findings

This part of the paper contains a brief description of the two case-studies comprising such sections as: the background of the research company, its experience in international technology partnerships, the essence of co-operation projects and their inertia. Interviews were taken in 2007 and all the data of both the cases are traced back to that period.

Background of the First Case-Study: Piezo Group

A Piezo plant was established in 1944 by the Soviet Government. The objective was to provide with the piezoelectric resonators the radio systems of the Soviet army. With market reforms in Russia, the Soviet Piezo plant was restructured and on its ground was organised an Open Joint-Stock Group of Piezo Companies, known since 1995 as Piezo Group. Piezo Group specialises in R&D, manufacturing and delivering of piezoelectric monocrystals and products on the basis of crystal elements, piezoelectric sensors, quartz and LGS crystals, quartz filters and highly stable crystal oscillators. Areas of technologies applications are telephone; satellite and mobile communication; aircraft and space; oil-and-gas industry; navigation; home appliances; management devices of technological processes; computer, radio and automobile engineering. The production process is certificated by ISO 9001. In 2006 only, Piezo received 10 patents in the field of technology and design of new materials and products. Achievements of the R&D departments of Piezo Group are being regularly published in the Russian and international journals, presented at international symposiums, conferences, workshops, and exhibitions.

Experience in International Technology Partnership

Piezo Group has gained good experience in international partnerships. Foreign co-operation history of Piezo is coming back to as early as 1970s when the Soviet plant started its export activities. For today Piezo divides its international experience into two phases that are *prior* and *post* to the market reforms.

The first part of international collaboration covers the years of 1970-1990s. During those years Piezo plant exported technological equipment to countries belonging to socialist block, particularly, Bulgaria, Poland, and Czechoslovakia. During 1980s international co-operation flourished at Piezo most intensively, though, not independently. It was the time when the USSR Ministry of Foreign Commerce searched for partners for Piezo and supported at large international success of the plant. For the reasons of supervision, the Soviet Government provided international experts for Piezo's trips to Switzerland, America, Japan, etc. Those experts, working for the USSR Ministry of Foreign Commerce, used to specialise in different countries and were well aware of particularities of the regions they were competent in. Contemporary Chief Executive Science Officer of Piezo asserts that such kind of consulting assistance is really in need now. At present the enterprise is seeking for experts in the global technology market to manage international R&D projects of Piezo Group more professionally.

The second stage of Piezo's international experience began with the introduction of market economy in Russia when the plant had to acquire independent administration and start to manage all the activities without support from the Soviet Government. By 2007 Piezo Group counted its modern experience in technology collaboration to be as long as 12 years. The enterprise has succeeded in establishing partnerships with companies from the USA, Japan, France, Switzerland, and Germany. For the moment Piezo is involved in negotiations with China and Vietnam and has already signed some preliminary agreements. Piezo's experience in international technology partnership in post-Soviet period could be further subdivided into *during-reforms* (circa 1995-2000s) and *after-reforms* (since 2000s) periods.

After the introduction of the market economy and collapse of the Soviet Union, Piezo made its first attempts to find external partners independently, without the governmental support and supervision. Piezo focused externally because the domestic business was in the deep economic crisis and internally no quick recovery was expected. At the very beginning of the reforms Piezo's management was looking for establishing joint venture with foreign companies. Along with the deep crisis of the domestic industrial sector the main reason for those strategic intentions was the lack of complete practice in the market-driven environment. Piezo badly searched for acquiring marketing knowledge from potential western partners.

There are two reasons why the strategies of joint ventures with international companies are no longer urgent today. Firstly, the economic situation has driven to its recovering and Russian industrial enterprises make their orders with Piezo more often.

Secondly, Piezo Group has gained enough managerial strength and economical vigour for the last twelve years in administrative independence. In changed circumstances, contract agreements are proved to be the most efficient forms of partnership with international companies. At present, major forms of international co-operation at Piezo are outsourcing and research contracts which categorise as customer-supplier modes of technology partnerships (Narula, 2003a).

Russian industrial consumers used to consider foreign products to be of higher quality than those produced domestically. In b2b market the attitude to national enterprises has changed for the better stance towards Russian suppliers of technologies only recently. "*Russian industrial enterprises have lately started to consider seriously Russian suppliers of technologies. The situation in the domestic market is changing for the better. At Piezo, we are trying hard to gain good reputation on the domestic market. When a customer needs to improve this or that technological parameter, we are always eager to offer our R&D solutions. The national market is the main market for us. It is tremendous in terms of the future sales and potential demand*". In the domestic market, Piezo has just signed a number of agreements with the Russian companies. These R&D projects are mostly focused on the substitution of Japanese and western technologies. Piezo has nearly 40 enterprises which used to work with imported technologies but recently has changed over to domestic (Piezo Group) technology innovations. Piezo Group improves technical parameters and substitutes imported technologies by its own and thus gain the niche on the Russian market. Internal market is enormous and by 2007 Piezo Group has reached those levels of sales that had been in the 1980s, before the market reforms started. In its strategic intentions on developing external technology partnerships, Piezo Group is less interested in international R&D than in the national domain because 90% of sales are done on the Russian market and it is far away from its saturation. At the same time, the management of the company considers it necessary to strengthen international technology co-operation and this is an indisputable issue for the Piezo Group.

International Technology Co-Operation and Inertia in It

At Piezo the process of international co-operation with foreign companies is organised as follows. As a new state of things, a foreign company initiates first contacts with Piezo and the opening round of negotiations starts. The core topic for discussion at this stage is the objective of R&D to be carried out. When the goals of R&D are clear, companies sign a research agreement. Then partners schedule the R&D plan. After the implementation of each stage of the research, Piezo forwards reports to its international counterpart. Piezo Group generally carries out small-scale international R&D projects in the field of piezotechnics with specific rather than strategic objectives with partners from France, Germany, or Japan. To add to R&D contracts, manufacturing plants belongings to Piezo Group sometimes export finished goods in small batches.

Piezo's experience in international co-operation shows that the terms and conditions of implementation of the projects are the major points for establishing partnership with foreign companies. "*There is no excuse for not accomplishing the project within the time stipulated in the contract*". It would "*cause the loss of reputation*". Force-major circumstances, of course, are taken into the contracts but the best way to avoid such situations is to "*have 10% of a time stock. As it is impossible to foresight many of the unexpected changes, the reasonable thing is to specify a longer period than it is really needed*".

Comparative analyses of geography in international technology co-operation attest that in establishing partnership Piezo encounters the greatest complexities with eastern companies: Japanese, Chinese, and Vietnamese. The primary problem is the lack of technical language interpreters. During negotiations Piezo R&D staff often has to discuss technical issues of the project using "*broken English language*" what greatly complicates business communication. "*When Vietnamese came to visit Piezo for negotiations, they even did not take an interpreter with!*" Collaboration of Piezo with Chinese companies is being developed under the scenarios which pass many international companies in China (Farhang, 1997). Piezo had around fifteen trips to China, meetings with large delegations on the level of provincial authorities, however, business cooperation or economic relationships with the Chinese have not yet been far progressed. For the moment Piezo has one patent issued in China.

The most successful international relationship Piezo has been developed with the countries from the EC, especially, with France and Germany. "*Unlike Chinese, Europeans sign agreements after 2-3 rounds of negotiations*". Piezo's management perceives that the paradigm of open innovation in the global R&D market has started its development only recently. A vivid example of continuing policy of "closeness" is the fact that representatives of some foreign companies could not share business cards. "*Eastern companies conduct many negotiations, arrange meetings and business dinners, speak about big projects, but further on such interactions hardly move ahead. Western companies also adhere to policy of restraint, especially, in regard to financial issues. Discussing their customers, they, as a rule, name 2-3 companies they work with – which are widely-known. Chinese are extremely 'closed'. They did not even give business cards sometimes – an elementary act for communication!*"

By the time passing, Piezo accumulates experience in the global R&D market. Staff gets greater confidence in negotiating and in formulating marketing demands. "*Herein operates the same principle as in the labour market: no experience – no job*". Gaining experience R&D experts of Piezo say with determination that the constituent attribute of any decision-making on technology partnership are the perspectives of further development of innovation. It is an essential part at any technical negotiations. "*Sometimes ways of improving technology become more important, than ways of patenting and protection of IPR. It concerns, of course, those technologies which in the long-term could easily lose their competitiveness. Even if we have a patent, international partners would discuss with us the opportunities of technology improvements*". While making a decision on a technology partnership, the questions on IPR being discussed in details, however, as the experience of Piezo confirms, the issues of incremental innovations are still more important than IPR. When negotiating IPR of the co-operative project, know-how is the major point for making the decision on its adoption. In international co-operation projects Piezo always keeps a part of know-how which they did not describe in the patents. This know-how becomes the core part for technology agreements. There is an interesting fact, Piezo likes to mention. "*When we received the European patent, which was issued by Geneva's Office, a number of foreign companies offered us to sign with them agreements of becoming our representatives in Europe*".

Apart from the domestic market (Tyler and Steensma, 1995), for international agreements the technology is more attractive at later stages of development as the results are to be seen in the very near future. "*It is well-known, that R&D phase is always shorter, than the commercialisation part of the innovation project and the organisation of production, when the very latter takes 40% of the whole time*". According to interviewees, Russian R&D staff in the field of piezo-technologies is in the demand on the global market for their high qualification rather the low payment (Fowles and Clark, 2005). Foreign companies collaborate with Russian R&D staff for their expertise and research data they could get from Piezo. "*High qualification of our R&D staff – is the cornerstone for international co-operation. Foreign companies value the significance of the data they could have from our technology experts*". At the same time, it is worth mentioning that the cost of the project carried out with Russian R&D staff remains low in comparison with global prices. Though, for Russian domestic S&T market the cost of technology development would still be rather high.

Today, Piezo Group does not encounter problems with searching for customers or findings partners to compare with the beginning of 1990s. Domestic and foreign companies often make orders and "*internet is of a great support here*". Piezo website is updated regularly and potential clients learn about its innovations and trace them through the net. Piezo confirms that nowadays a lot of international companies come to Russia for the initial contacts with their company rather than Piezo does it itself. Additionally, Piezo notices certain interest of foreign companies in the published works of scientific personnel of the Russian company. R&D staff is being constantly invited to participate in international symposiums held in America and is welcomed to publish research results in world scientific journals.

The sample of Piezo Group is the vivid example of an ex-USSR enterprise, originated under the command closed administrative system, and now being adjusted to the market economy and open innovation policy. Interestingly, in its external strategies Piezo Group has reoriented from joint venture perspectives to customer-supplier co-operative modes. Also for the study at hand, Piezo brings some comparison of western versus eastern technology partnerships and some evidence of their transition from the planned to the market economy.

Background of the Second Case-Study: Inkomet

In 1989 inside the closed administrative formation Sarov, known as Nuclear Center Arzamas-16, was established a private company Binar. The company was opened by R&D staff of the Russian Federal Nuclear Centre. Binar's spheres of activity are high-tech projects in electronic industry, software, and instrument engineering. Binar specialises in R&D, commercialisation of technology and product innovations, small-scale manufacture of electronic devices, development of software for the gas, oil and chemical industries and water transport. Binar's strategy is to cover the whole technology cycle including R&D, marketing research, production, and commercialisation. In 2004, on the basis of the privately owned parent company Binar, a new firm Inkomet started up. Inkomet's activities are R&D of the measurement instruments in the iron-and-steel manufacturing. The application of R&D results is in the planeness of metal-rolling. The start-up company created the best in the world contactless, opto-electronic measuring device. It became possible due to the joined R&D efforts of Sarov experts in several scientific fields. Particularly, R&D carried out in the *optical* field was applied to the *iron-and-steel* production. The cornerstone of Inkomet's success is the highest qualification of R&D staff of the Russian Federal Nuclear Centre (DTI, 2006). Despite the fact that Inkomet has been formed rather recently, it has already gained some experience in international technology partnership. International history started for

Inkomet in 2006, when a company from Germany offered the Russian start-up to organise a joint venture.

Experience in International Technology Partnership

The German company, working in the same field, suggested Inkomet joint commercialising of the new technology device in the global market. The Germans has representatives in 30 countries what makes the offer of technology partnership especially interesting for the Russians. Inkomet is aware that without the German partner they are unlikely to enter the global market. To make the same or similar distribution network as the Germany company has succeeded in – "*would take us 10 or 15 or even more years*". The distribution network would require a lot of investment that are difficult to get. By that time the device would become obsolete and lose its competitiveness. When the joint venture was established, Russians realised their dependence on Germans. Foreign partners perceiving it as well "*dictated terms and conditions of technology co-operation*". The case of Inkomet is valuable for the study at hand as it comprises the creation of the joint venture between a Russian company from a closed atomic town and a company from a developed economy. The importance of this case is also in the fact that the German partner was first to start negotiations. Additionally, the case of joint venture includes all the stages of cooperation: R&D, manufacturing and sales.

International Technology Co-Operation and Inertia in It

An interesting fact for the Russian-German collaboration is that the company from Germany made the first steps towards the initiation of the joint venture. More than that, they have been following the activity of Sarov's firm for some period of time. German colleagues often came to presentation, which Russian engineers arranged at domestic factories to display the innovative device for metallurgists. Inkomet used to meet their future partners at exhibitions where they also presented their innovative technology. Publications and two patents "*could have also influenced upon the intention of the foreign firm to enter into the agreement with the Russian company*". The lack of experience of Inkomet in international technology cooperation has not stopped the German company. "*For a partnership with a foreign company, the uniqueness of the technology is of the utmost importance*".

Despite the fact that German company specialises in the field of measuring devices and makes similar equipment, the Germans are not interested in purchasing the technology and, as the major condition of the partnership, have put forward the organisation of the joint venture. This is another remarkable detail of the case. "*If it is the question of an innovative technology the German company sees cooperation only in the creation of a joint venture. They motivate such a standpoint by the necessity of accompanying innovation and its further improvement. It is true that sometimes no one can do it better than the authors of the innovative idea. And if the technology is not incremented, ahead or together with the market needs, it becomes obsolete extremely soon*".

Other striking feature of the studied Russian-German partnership is the absence of cultural and language barriers between the partners. Such a favourable situation for this international project is possible through a participation of a German citizen with the Russian background. It happened so, that one of the employees of the German company is the former Russian citizen who had been living in Russia till his 35 years old. Then he emigrated from Russia to Germany. It is no wonder, he knows the culture of both countries and this circumstance embeds in the better communication relations of the Russian-German partnership.

At the moment the Russian company and its German partner are working on the scheme of their co-operation which is to schedule the major elements of managerial coordination and organisational interactions in the field of production, sales and costs. The thing is that the manufacturing of the Russian equipment depends on several constituent elements. It is planned that the place of assembly would be in Germany. To lower costs of the joint venture, a group of manufacturing constituents which could not be replaced by the German analogues, would be made in Russia. Those parts of the assemblies that could be bought in the

EC would be delivered from Europe. The German side will make the production and the sales worldwide while their Russian counterpart is responsible for technical improvements and further development of the innovative device.

The German company fully exploits the lack of the marketing experience of the Russian innovators in commercialising new technologies, "*putting every possible pressure*". The German party always receives advantages in conditions of cooperation. Inkomet assumes that if they have had enough experience in commercialising innovations internationally, they would behave differently. Experience adds much to confidence. "*If we have managed to take part in let say 10 co-operative projects, of course, we could show more assurance during the negotiations. We would not feel the difference with our foreign colleagues. We would be able to be more persistent, could defend our interests, might direct our terms and conditions and would not obey to everything we are dictated by our foreign partners*".

The core reason for making technology partnership with the Russian company is the quality of the new technological equipment and its final price on the global market in comparison with foreign analogues. The most attractive aspect for foreign companies as the low labour cost, specific for the Russian R&D market in comparison with some other countries, is not the reason for the partnership of Sarov start-up. "*The price as we would wish it – is no use to negotiate about. It is important to understand, that the price is set by the market. If to request a high price for our device nobody will buy it. The innovative product should not only be better than analogues, but also cheaper*". The cooperation of Sarov firm with the foreign partner is based on the scheme that the technology is not directly sold to the German company. End users are metallurgical plants all around the world. That is why both the parties are interested in the competitiveness of their joint venture; therefore both try to lower costs.

As for the issues of IPRs of the new technology, the experience of Inkomet shows some inertia here. In the first place, IPRs are the basis of any negotiations. Sarov engineers evidenced that "*nobody will talk to you if the rights are not protected and if the IP does not belong to you. Whether the technology is protected only in Russia or somewhere abroad as well – is another group of questions for negotiations. The fact of protection is of the utmost importance, as for 'how and when' it is the theme for further interactions. Herein, we always have to make a difficult choice between patenting the technology or know-how. On the one hand, a patent protection requires the opening of the technology. On the other hand, patenting is gaining particular importance in the globalisation of innovations and internationalisation of R&D. If we won't protect the technology, it could be easily taken away, even by our potential partners. The problem of many of the Russian R&D organisations is the lack of experience in IPRs management – we often miss the period of foreign patenting. Most likely, the know-how protection is to be recognised as one of the best ways of IPR protection".*

As a matter of fact, Inkomet considers internal customer-supplier relations of more value than external due to the increase of the orders from domestic metallurgical plants. Managers of Inkomet explain their standpoint through open innovation paradigm. The facts that foreign companies are interested in external suppliers and that international companies come to Russia in the search for new solutions are regarded as the signs of "openness". "*Openness of the foreign companies is evidenced by initiating contacts, their will and readiness for cooperation, aspiration for realisation of innovation projects by joint efforts. Along with some attributes of "openness", western companies deal with Russian companies in a special way. For instance, our practice shows that foreign companies do not trust us in strategic issues, meaning that they do not treat us as their equals in term of the marketing knowledge and consumer understanding. I believe that the origin of technology imposes certain traits on processes of interaction between us".*

Despite of all challenges of working with foreign partners, Inkomet is satisfied with the overall results, as well as with the marketing knowledge and commercialisation expertise they are gaining. The research sample of Inkomet is the book example when a start-up company finds a strategic partner investing into an innovation project through the creation of a joint venture. In case with Inkomet, the Russian technology is entering the global market via support of a large European company specialising in same area, possessing good reputation in the market, owning free investments, but not having a progressive technology.

Conclusion and Discussion

This research studies two cases of international technology partnership taking Russian R&D organisations for a sample. The co-operation interrelations are customer-supplier and joint-venture organisational modes. In terms of source, recipient, technology, management, and country-specific attributes each case delineates attributes that are (i) neutral, (ii) contributing, or (iii) hindering the establishment of international technology collaboration with Russian R&D organisations. Table 2 summarises in juxtaposition the results of two cases.

Table 2. A juxtaposition of the attributes and modes of international co-operation based on the studies of Russian R&D organizations

Factor	*Focus*	*Customer-supplier case-study of Piezo*			*Joint-venture case-study of Inkomet*		
		Contributing	*Neutral*	*Hindering*	*Contributing*	*Neutral*	*Hindering*
Source	Cooperative			–		~	
	history	+			+		
	Expertise level		~				–
	IP rights						
Recipient	Compatible		~			~	
	goals	+			+		
	Contribution		~				–
	Patterns of						
	payoffs						
Technology	Further		~				–
	development	+			+		
	Expected cost		~			~	
	Life cycle						
Management	Coordination	+			+		
	Mutual trust			–			–
	Mentality			–		~	
Country-	Communication		~		+		
specific	Cultural	+				~	
	differences			–			–
	Marketing						
	experience						

(1) To understand a *source* group of attributes, from theory and practice are drawn such criteria as *cooperative history, expertise level, and IPRs*. Tyler and Steensma (1995) argue that the cooperative history of the potential partner is a reliable measure for collaboration. They evidence that "the more favourable the potential partner's cooperative history the more attractive technological collaborative relationships will be to top executives, all else equal". The experiences of Piezo and especially of Inkomet attest that prior experience of collaboration is the factor which is to be concerned more by Russian R&D organisations rather than their foreign counterparts. The lack of experience of Russian R&D organisations is not strongly considered by international company as an impediment for establishing the partnership. On the contrary, the insufficiency of the previous practice is used by foreign company as a right to dictate terms and conditions of the technology co-operation.

For international companies the level of technological complexity and technical expertise of Russian R&D staff overvalue the shortage of external co-operative knowledge (Trifilova, 2007b). Until recently, Russia was governed by a Soviet model of the central planning which supported commanded resource allocation. In Soviet Russia the strength of the S&T sector was considered of great importance amongst the political objectives of the whole country and its international prestige. The Soviet Government created an extremely large R&D base, "greater, in absolute terms, than that of most of the industrially developed nations" (Gokhberg *et al.*, 2001, p. 4). By 1990 R&D personnel in the Soviet Russia exceeded 2.8 million (S&T Overview, 2006). Even considering the significant losses that have been happened in S&T sector after the USSR collapse, Russia is inferior only to Sweden, Japan, the United States, Finland, Switzerland and France in absolute value of the R&D personnel (Gokhberg *et al.*, 2001). The expertise of "human resources remains Russia's greatest asset…be[ing] the core of national S&T and innovation policy" (*ibidem).*

IPR is a question for careful consideration on technology partnership with Russian science-intensive and research institutes. IPRs are important because "unknown or unclear ownership creates uncertainty which acts as deterrent to investment" (Watkins, 2003, p. 19). As Watkins explains it "no rational individual would purchase a house without obtaining clear title, no rational investor would invest in agricultural improvements without clear title to land, and the situation with IP is analogous" (*ibidem*). IPR regulation is an inherent impediment provoked by the unresolved debate in Russia about IP created at the expense of the budget resources. As mentioned above, in the period of the Soviet Union all the R&D establishments in the country were funded by the government and all the inventions were the state property. Transition to the market economy and privatisation policy of the state property, raised a question between the government and research establishments who should own budget financed IP. The situation remains unresolved and for some Russian science organisations it becomes an impediment for the process of commercialisation technologies on domestic and especially international market. The results of the cases studied evidence that Russian R&D organisations encounter problems with IP administration due to the lack of experience in managing innovation projects independently, without the governmental support.

Summarily for Russian case-studies, of three attributes in the *source* group: (1) cooperative history is suggested to be a neutral factor; (2) expertise level is claimed to be a part contributing to the establishment of international technology partnership; and (3) the issue of IPRs is regarded to be a problem hindering the establishment of external partnerships.

(2) To understand a *recipient* group of attributes, from theory and practice are drawn such criteria as *compatible goals, contribution and patterns of payoffs*. Both Piezo and Inkomet attest that they are more interested in domestic rather than international market. Notwithstanding, both the Russian companies participate in international cooperation; they understand strategic importance of international co-operation for their competitiveness on the national domain. Technological data is the major contribution of Russian R&D organisations. Their foreign counterparts contribute marketing knowledge and commercialisation strategies.

Spekman *et al.* (1997) recognise that dependence may motivate collaboration, though working consensus is more necessary to facilitate co-operation. Partnerships exhibiting a working consensus will be those in which both parties believe that their own investments are substantial and each recognises the valuable endowments of the other. In the recipient group of attributes patterns of payoffs appear to be a disputable question. The case of Piezo and Inkomet delineate some contradiction in the perspectives and the actual state of things. Interviewees seem to fall short of expectation with the economic return of the projects. They were only secured about management and market knowledge they are gaining from international partners.

Summarily for Russian case-studies, of three attributes in the *recipient* group: (1) compatible goals is suggested to be a neutral factor; (2) contribution is claimed to be a part supporting to the establishment of international technology partnership; and (3) patterns of payoffs are regarded to be a problem that hindering Russian R&D organisations from establishing external partnerships.

(3) To understand a *technology* group of attributes, from theory and practice are drawn criteria of *further development, expected cost, and life cycle.* By technology is broadly understood know-how; techniques, patented or otherwise proprietary processed; materials; equipment; systems etc. (Siegel *et al.*, 1995). Many researchers use the definition of Abetti (1989) who defines technology as "a body of knowledge, tools, and techniques, derived from both science and practical experience, that is used in the development, design, production, and application of products, processes, systems, and services" (cited from Tyler and Steensma, 1998, p. 941). It means that technology not only entails new product innovations and new processes by which outputs are generated, but also includes scientific and practical know-how that comes with experience. Technology often refers to new and better ways of achieving economic ends that contribute to economic development (Phillips *et al.*, 1994).

Both of the studied companies assure that for their international partners the issues of further development of a technology and potential ways of improvement are more important than IPRs or even costs. According to Minford *et al.* (1997), when a country emerges it means that it enjoys: a high rate of inward investment; fast growth, primarily in manufacturing; and low labour costs. Russia, belonging to emerging market economies, is by definition notable for low labour costs. An American software company with development resources in the Urals region of Central part of Russia is quoted as saying that each Russian staff member can replace four American staff members but costs one quarter of an American staff member (Global Watch, 2006). The results of the study suggest that high quality of R&D data rather than low labour price is contributing to international technology collaboration. Costs, as assured by Inkomet, are of a more concern for a joint venture co-operative mode.

Tyler and Steensma (1995) argue that for cooperative relationship the earlier the stage of technological development, the better. In maturity, when market uncertainties are resolved, acquisitions and licensing may be preferred. At the same time, early stages require more exchange of information and interrelation, what seems rather problematic for international collaborative projects when the partners are in different countries or parts of the world. The results of the research reveal that Russian science-intensive enterprises find partners for commercialisation of technology at later stages.

Summarily for Russian case-studies, of three attributes in the *technology* group: (1) further development is regarded to be an impediment that retards the initiation of external partnership; (2) expected costs, which are assumed to be low in comparison with the global price, are claimed to be a part supporting to the establishment of international partnership; and (3) life cycle is suggested to be rather a neutral factor.

(4) To understand a *management* group of attributes, from theory and practice are drawn such criteria as *coordination, mutual trust, and mentality.* Albino *et al.* (2004) study the process of knowledge transfer as consisting of sender's cognitive system; recipients or user's cognitive system; codification; interpretation, and knowledge object. They split the model into two parts. The first describes the process through which knowledge flows from a source to a recipient. The second depicts the process through

which the recipient puts knowledge into practice. Knowledge transfer depends on the recipient's characteristics, such as experience, values, motivation, and beliefs. Cohen and Levinthal (1990) earlier summarised this idea by the absorptive capacity. They defined it as the ability of individuals in an organisation to assimilate and apply new information by relating it to prior existing knowledge. Shortage of necessary information is a problem general and hard to overcome for technology co-operation. The case of Piezo evidences that sharing of project information, especially providing the partner with research data in the scheduled time, are the cornerstones for customer-supplier relationship. Partners' mentality in terms of closed vs. open innovation paradigm or developed vs. emerging market economy, as delineated by the cases of Piezo and Inkomet, only adds to the overall environment of co-operation. As for mutual trust, Inkomet underlines its lack with their partners and Piezo fears to lose it through accidental lagging behind the schedule. The main approach to all the partnerships is "win-win-win deal for all parties", being "the best reason for wholehearted participation" (Stamm, 2004).

Summarily for Russian case-studies, of three attributes in the *management* group: (1) coordination is claimed to be a part contributing to the establishment of international technology partnership; (2) mentality is suggested to be rather a neutral factor; and (3) mutual trust is regarded to be a problem hindering the establishment of external partnerships.

(5) To understand a *country-specific* group of attributes, from theory and practice are drawn such criteria as *communication, cultural differences, and marketing experience*. Russian country-of-origin specifics in the area of S&T have been described by many scholars (Watkins, 2003; OECD, 2004; Gokhberg, 2004; Gokhberg and Shuvalova, 2004; Lachinov, 2005; S&T Overview, 2006). The major weakness of the Russian S&T sector is its low inexperience in marketing own innovations. The majority of R&D establishments suffer from a deficiency of staff experienced in managing international technology collaboration. The real strength of the Russian S&T sector is its highly qualified technical personnel, greatly experienced in making technological breakthrough. Such

contradiction could be traced back to the ex-USSR R&D system when the Soviet government, considering science as its international prestige, devoted resources to creating new technologies paying little attention to economic return. Working in closed research institutes or isolated atomic cities Russian R&D personnel have gained a unique experience in many scientific fields but not in marketing knowledge.

In the cases studied, Piezo suggests that the lack of interpreters could block interactions with international partners, especially with the eastern group of countries. The case of Inkomet, on the contrary, shows the absence of this problem thanks to bilingual employee. One of the traits of the current globalization is the migration of people and workforce. The case of Inkomet attests that if to hire bi-national personnel, cultural differences could cause no problem apart from the marketing culture.

Summarily for Russian case-studies, of three attributes in the *country-specific* group: (1) open communication, mostly language communication, is claimed to be a part contributing to the establishment of international technology partnership, all else equal; (2) cultural differences are suggested to be regarded as rather a neutral factor contingent on the part of the world the partner is from; and (3) marketing experience is considered to be a problem hindering the establishment of external partnership.

The findings of this research shed some light on the implications for the management of international technology collaboration with Russian R&D organisations. First, the paper brings the evidence that Russian R&D organisation have overcame the transition from the command to the market economy and are adjusting to new environment. Second, the study gains some insight into the reasons and motives for international technology partnership with Russian R&D organisations. Third, the research highlights some neutral, contributing and hindering attributes of technology co-operation with ex-USSR R&D organisations. Forth, it calls for different management attention to the modes of collaboration.

Although the research results are interesting, limitations exist. An inherent limitation is that it examines two positive cases of international partnerships. In the analysis the study relies only on the respondents of the Russian R&D organisations which may have a biased view of what is actually occurring. Future studies could conduct investigations of the incomes and outcomes of establishing international technology partnerships with Russian R&D organisations. This could be achieved applying resource-based theory and taking for a research sample both Russian and non-Russian counterparts of technology co-operation projects.

References

Albino, V, Garavelli, A, Gorgogolione, M (2004). Organisational and technology in knowledge transfer. *Benchmarking: An International Journal*, 11(6), 584–600.

Alcorta, L, Plonsky, G, and Rimoli, C (1998). The experience of technological collaboration by MERCOSUR companies. *Technology Analysis & Strategic Management*, 10(3), 341–363.

Allio, R (2004). CEO interview: the Innocentive model of open innovation, *Strategy & Leadership*, 32(4), 4–9.

Belderbos, R (2003). Entry mode, organisational learning, and R&D in foreign affiliates: evidence from Japanese firms. *Strategic Management Journal*, 24, 235–259.

Benedetto, A, Calantone, R and Zhang, C (2003). International technology transfer. Model and exploratory study in the People's Republic of China. *International Marketing review*, 20(4), 446–462.

Bennett, D, Hongyu, Z, Vaidya, K and Ming, W (1997). Transferring manufacturing technology to China: supplier perceptions and acquirer expectations. *Integrated Manufacturing Systems*, 8(5), 283–291.

Brooking, A (1996). *Intellectual Capital.* International Thomson Business Press: London.

Bruce, M, Leverick, F, Littler, D and Wilson, D (1995). Success factors for collaborative product development: a study of suppliers of information and communication technology. *R&D Management*, 25 (1), 33–44.

Carneiro, A (2006). Adopting new technologies. In *Handbook of Business Strategy*, pp. 307 – 312.

Chesbrough, H (2003). *Open Innovation. The New Imperative for Creating and Profiting from Technology.* Harvard Business School Press: Boston.

Chiesa, V, Gilardoni, E and Manzini, R (2005). The valuation of technology in buy-co-operate-sell decisions. *European Journal of Innovation Management*, 8(2), 157–181.

Chiesa, V (1996). Evolutionary patterns in international research and development. *Integrated Manufacturing Systems*, 7 (2), 5–15.

Cohen, WM and Levinthal, DA (1990). Absorptive capacity: a new perspective on learning and innovation. *Administrative science Quarterly*, 30, 128–152.

Czuchry, A and Yasin, M (1999). The three "Is" of effective marketing of technical innovation: a framework for implementation. *Marketing Intelligence*, 17(5), 240–247.

Dalton, D, Serapio, M and Yoshida, P (1999). *The globalisation of industrial R&D*. US Department of Commerce, Washington, DC: NTIS.

Das, TK and Bing-Sheng, T (1998). Between trust and control: Developing confidence in partner co-operation alliances. *The Academy of Management Review*, 23(3), 491–513.

Dodgson, M (1993). Learning, trust, and technological collaboration. *Human Relations*, 46(1), 77–96.

DTI (2006). *The Global Partnership: Fourth Annual report*. DTI: Ministry of Defence.

Ettlie, J (2006). *Managing innovation*. Elsevier: USA.

Farhang, M (1997). Managing technology transfer to China. Conceptual framework and operational guidelines. *International Marketing Review*, 14(2), 92–106.

Farr, CM and Fisher, AW (1992). Managing international high technology cooperative projects. *R&D Management*, 22 (1), 55–67.

Ferguson, D (2005). Partnering for innovation and growth. In *Handbook of Business Strategy*, pp. 101–106.

Fowles, S and Clark, W (2005). Innovation and networks: good ideas from everywhere in the world. *Strategy and Leadership*, 33(4), 46–50.

Gokhberg, L (2004). Russia's national innovation system and the "New Economy". *Problems of Economic Transition*, 46(9), 8–34.

Gokhberg, L, Kovaleva, N and Kouznetsova, I (2001). *Innovation management in Russia: a review of training needs and opportunities for growth*. British Council: Russia.

Gokhberg, L and Shulanova, O (2004). *Russian public opinion of the knowledge economy: science, innovation, information technology and education as drivers of economic growth and quality of life*. British Council: Russia.

Gunasekaran, A (1998). An analysis and experience on the management of international and joint R&D projects. *Management Decision*, 66(10), 669–678.

Hagedoorn, J (1993). Understanding the rationale of strategic technology partnering: interorganisational modes of cooperation and sectoral differences. *Strategic Management Journal*, 14, 371–385.

Hakanson, L (1993). Managing cooperative research and development: partner selection and contact design. *R&D Management*, 23 (4), 273–285.

Hagedoorn, J, Sedaitis, J (1998). Partnerships in transitional economies: international strategic technology alliances. *Research Policy*, 27, 177–185.

Harris, RC, Insinga, RC, Morone, J and Werle, M (1996). The virtual R&D Laboratory. *Research Technology Management*, 39(2), 32–37.

Hausler, J, Hohn, HW and Lutz, S (1994). Contingencies of innovative networks: A case study of successful inter-firm R&D collaboration. *Research Policy*, 23, 47–66.

Hoetker, G (2005). How much you know versus how well I know you: selecting a supplier for a technically innovative component. *Strategic Management Journal*, 26, 75–96.

Horton, V (1998). An exploration of strategic collaboration in the Triad. *European Business Review*, 98(1), 4–12.

Humphrey, J and Schmitz, H (1998). Trust and Inter-firm relationships in developing and transitional economies. *The Journal of Development Studies*, 34 (4), 32–52.

Kayes, DC, Kayes, AB and Yamazaki, Y (2005). Essential competencies for cross-cultural knowledge absorption. *Journal of managerial Psychology*, 20(7), 578–589.

Kim, Y and Lee, K (2003). Technological collaboration in the Korean electronic parts industry: patterns and key success factors. *R&D Management*, 33 (1), 59–77.

Kleinsmann, M and Valkenburg, R (2005). Learning from collaborative new product development. *Journal of Workplace Learning*, 17(3), 146–156.

Kogut, B (1988). Joint ventures: theoretical and empirical perspectives. *Strategic Management Journal*, 9, 319–332.

Lachinov, M (2005). Russia: science and technology framework, accessed from www.alwatchservice.com. (14 April 2005).

Lin, C, Tan, B and Chang, S (2002). The critical factors for technology absorptive capacity. *Industrial Management and Data Systems*, 102(6), 300–308.

MacDougall, S and Hurst, D (2005). Identifying tangible costs, benefits and risks of an investment in intellectual capital. *Journal of Intellectual Capital*, 6(1), 53–71.

Matthews, J (2006). Microwave power: Russia's strength. *Global Watch Magazine*, 6, 10–11.

Ming, W and Xing, Z (1999). A new strategy of technology transfer to China. *International Journal of Operations & Production,* 19(5/6), 527 – 537.

Mohr, J and Spekman, R (1994). Characteristics of partnership attributes, communication behaviour and conflict resolution techniques. *Strategic Management Journal*, 15(1), 135–152.

More, E, and McGrath, M (1999). Working cooperatively in an age of deregulation. *The Journal of Management Development*, 18(3), 227–254.

Narula, R (2003a). Globalisation and trends in international R&D alliances. *MERIT-Infonomics Research Memorandum series*, 2003-001.

Narula, R (2003b). The international dimension of innovation. *MERIT-Infonomics Research Memorandum series*, 2003-009.

Narula, R (2004). Foreign direct investment through acquisitions and implications for technological upgrading: case evidence from Tanzania. *MERIT-Infonomics Research Memorandum series*, 2004-008.

Narula, R (2005). Using multi-hub structures for international R&D: organisational inertia and the challenges of implementation. *MERIT-Infonomics Research Memorandum series*, 2005-025.

Nobel, R and Birkinshaw, J (1998). Innovation in multinational corporations: control and communication patterns in international R&G operations. *Strategic Management Journal*, 19, 479–496.

OECD (2004). Science, technology and industry outlook. Country response to policy questionnaire. Russia.

Parker, H (2000). Interfirm collaboration and the new product development process. *Industrial Management and Data System*, 100(6), 255–260.

Parkhe, A (1993). Strategic alliances structuring: a game theoretic and transaction cost examination of interfirm cooperation. *Strategic Management Journal*, 36(4), 794–829.

Piachaud, B (2005). Outsourcing technology. *Research Technology Management*, 48(3), 40–47.

Poulymenakou, A and Prasopoulou, E (2004). Adopting a lifecycle perspective to explain the transition from technological innovation to alliance management. *Management Decision*, 42(9), 1115–1130.

Powell, S (2005). Ramifications of globalisation. Michael Treacy in conversation with Sarah Powell.

Radovilsky, Z (1994). Managing Operations in the Former Soviet Union: Current Situation and Future Development. *International Journal of Operations and Production Management*, 14(2), 43–50.

Ragatz, GL, Handfield, RB and Scannell, TV (1997). Success factors for integrating suppliers into new product development. *Journal of Product Innovation Management*, 14, 190–202.

Rothaermell, F and Deeds, D (2004). Exploration and exploitation alliances in biotechnology: a system of new product development. *Strategic Management Journal*, 25, 201–221.

Sakakibara, K (1993). R&D cooperation among competitors: a study of the VLSI semiconductor research project in Japan. *Journal of Engineering Technology Management*, 10, 393–407.

S&T, Overview (2006). Russia. www.globalwatchservice.com/Pages/Russia/S&T database [5 January 2006].

Saxton, T (1997). The effects of partner and relationship characteristics on alliance outcomes. *Academy of Management Journal*, 40(2), 443–461.

Serapio, M and Hayashi, T (eds.) (2004). *Internalization of Research and Development and the Emergence of Global R&D Networks*. Elsevier: Oxford, UK.

Souder, WE and Nasser, S (1990). Managing R&D consortia for success. *Research Technology Management*, 33(5), 44–50.

Spencer, J (2003). Firms' knowledge-sharing strategies in the global innovation system: empirical evidence from the flat panel display industry. *Strategic Management Journal*, 24, 217–233.

von Stamm, B (2004). Collaboration with other firms and customers: innovation's secret weapon. *Strategy and Leadership*, 32 (3), 16–20.

Swan, JA and Newell, S (1994). Manager's beliefs about factors affecting the adoption of technological innovation. *Journal of Management Psychology*, 9(2), 3–11.

Teece, DJ (2000). *Managing Intellectual Capital*. Oxford University Press.

Teichert, T (1993). The success potential of international R&D cooperation. *Technovation*, 13 (8), 519–532.

Todeva, E, and Knoke, D (2005). Strategic alliances and models of collaboration. *Management Decision*, 43(1), 123–148.

Trifilova, A (2006). Russian science and technology: A case study of the Volga-Vyatsky region. In *Proc. 8th Int. GBATA Conf.* Management and Technology in the Global Economy: Nurturing Innovations and National Heritage. Moscow: Russia.

Trifilova, A (2007a). Geography of technological co-operation: the Russian study. In *Proc. 18th Int. ISPIM Conf.* Innovation for Growth: The Challenges for East & West. Warsaw: Poland.

Trifilova, A (2007b). Is the origin of a technology an inherent risk or an externally generated uncertainty for R&D managers? In *Proc. 30th Int. RADMA Conf.* Risk and uncertainty in R&D management. Bremen: Germany.

Trifilova, A, Bartlett, D and Altman, Y (2007). Innovation and Technological Collaboration: a Russian perspective. In *Proc. 21st Int. BAM Conf.* Managing Research, Education and

Business Success: is the future as clear as the past? Coventry: UK.

Tyler, B and Steensma, K (1995). Evaluating technological collaborative opportunities: a cognitive modelling perspective. *Strategic Management Journal*, 16, 43–70.

Tyler, B and Steensma, H (1998). The effects of executives' experiences and perceptions on their assessment of potential technological alliances. *Strategic Management Journal*, 19, 939–965.

Watkins, A (2003). F*rom knowledge to wealth: transforming Russian science and technology for a modern knowledge economy*. World Bank.

Yin, R (2003). *Case Study Research: Design and Methods*. Thousand Oaks: Sage.

Zain, M and Yasin, N (1997). The importance of country-of-origin information and perceived product quality in Uzbekistan. *International Journal of Retail & Distribution Management*, 25(4), 138–145.

Zander, I, Solvell, O (2004). Cross-border innovation in the modern multinational – three dilemmas. In Serapio, M and Hayashi, T. (eds), *Internalization of Research and Development and the Emergence of Global R&D Networks'*, Elsevier, Oxford, UK.

PART 3

" Services "

- *Building a global electronic manufacturing service provider : vendor's process in outsourcing*
- *support services in supporting B-TO-B relationships*
- *Designing Rapid Services for Competitive Manufacturing*

BUILDING A GLOBAL ELECTRONIC MANUFACTURING SERVICE PROVIDER: VENDOR'S PROCESS IN OUTSOURCING

ZORAN PERUNOVIĆ

Technical University of Denmark,
DTU Executive School of Business, Building 421
2800 Kongens Lyngby, Denmark
Telephone: +45 45 25 61 15 – Fax: +45 45 88 43 37 – E-mail: zope@business.dtu.dk

MADS CHRISTOFFERSEN

Technical University of Denmark,
DTU Executive School of Business, Building 421
2800 Kongens Lyngby, Denmark
Telephone: +45 45 25 61 15 – Fax: +45 45 88 43 37 – E-mail: mchr@business.dtu.dk

Abstract

The aim of this paper is to make a contribution to the mapping of the vendor's process in outsourcing and to explore how that process has been built over the time. In addition, the paper aims at coming to grips with the fundamental issue of how an electronic manufacturing service (EMS) provider can develop its competences and capabilities to compete effectively and efficiently in this increasingly more competitive industry. The paper has been written in a teaching case style where the case of Escatec, a tier-two EMS has been presented. Escatec, have been adding various skills, knowledge and technologies into a set of competences that have been integrated into a well-structured process that reflects customers' requirements. Findings from the case suggest that the vendor's process is cyclic in its nature, with equally strong emphasis on getting the new and retaining the existing customers. Based on the analysis of Escatec's process and competences we propose that the key success factors for a successful one-stop-shop type of EMS are to 1) Establish rich portfolio of competences; 2) Integrate the competences into a process; 3) Ensure that process depicts the nature of the industry; 4) Develop a modular process that will enable capturing of customers anywhere along the value chain; 5) Continuously improve competences that mostly contribute to the organization's performance.

Key words: Outsourcing, EMS, vendor's process, competences

Inspiration and Problem Statement

The outsourcing phenomenon has been increasingly receiving attention both from academic and practitioners' communities. Several outsourcing process frameworks have been developed to describe the outsourcing-related activities of an organization. However, seeing the outsourcing process from the perspective of the outsourcer can give a biased picture of the outsourcing phenomenon. The vendor is as critical to the success of outsourcing as the outsourcing company is.

The body of knowledge on vendor's perspective in outsourcing contains discussion of the advantages and disadvantages of a company's engagement in an outsourcing arrangement, (Rochester and Rochester, 1995), approaches to customers, contracts and relationships (Mohr and Spekman, 1994; Willcocks and Choi, 1995; Pinnington and Woolcock, 1997; Zviran et al., 2001; Kern et al., 2002; Cullen and Willcocks, 2003; Liang and Xue, 2004; Blois, 2004; Stjernström and Bengtsson, 2004; Click and Duening, 2005; Oza et al., 2006), vendor's value proposition (Levina and Ross, 2003), vendor's capabilities necessary to win and run outsourcing contracts (Feeny et al., 2005), and in some scattered fragments, structure of the vendor's outsourcing process (Jenster et al., 2005; Vollman and Condor, 2000). Seen from the vendor's perspective, the most important three stages of an outsourcing process are to win, run and renew outsourcing contracts.

A quick overview on the topics within the vendor's perspective in outsourcing reveals that there is a significant deficiency of studies with the aim of structuring the vendor's process in outsourcing. Therefore the aim of this paper is to make a contribution to the mapping of the vendor's process in outsourcing and to explore how that process has been built over the time.

Success of an outsourcing arrangement is determined by numerous factors, where vendor's capabilities and competences seem to be some of the most important. For instance, Levina and Ross (2003) showed that vendors' capabilities have been recognised as the most important factors for the success of outsourcing. According to them, vendor's capabilities comprise its technical competence, understanding the customer's business, and relationship management. These capabilities and competences can be perceived as generic and applicable across industries. However, one has to be careful because for different industries one may expect to find different capabilities and competences that count for winning and renewing the outsourcing contract.

Within this context the paper aims at coming to grips with the fundamental issue of how an electronic manufacturing service (EMS) provider can develop its competences and capabilities to compete effectively and efficiently in this increasingly more competitive industry. Some of the questions we are pursuing in this paper are related to how can an EMS organize its human/organizational resources and internal processes in order to meet existing customers' need and how should it further develop its capabilities adapt to the changing conditions of the market place and thereby acquire new customers?

The paper is built on a detailed, specific case study of a multinational EMS company Escatec, headquartered in Malaysia[11]. In the next section we describe the sites' visits and the process of writing the teaching case. We proceed with setting the stage for the case i.e. describing the characteristics and recent developments on the EMS industry. Afterwards we introduce Escatec as a tier-two EMS provider and describe how they have been extending the portfolio of services/competences and how they have been integrated into the EMS process. We conclude the paper by presenting the major findings and implications for academia and industry.

[11] The paper is based on the research material used to develop case study **"Escatec: Building a Global EMS Company"**. The case is registered at ECCH under the reference # 308-044-1

Methodology

The field work in Escatec started in November 2006 when the first author visited headquarters in Malaysia. That visit had lasted for four working days. The same site was visited in August (two days) and November (four days) 2007. The Swiss site was visited in April 2007. Both authors were present during the interviews in April 2007 in Switzerland and in August 2007 in Malaysia. During those four visits we have talked to many informants in Escatec. On the corporate level we have interviewed the Executive Chairman, Chief Operating Officer, Corporate IT Manager, Corporate Business Development Director, Corporate Procurement Manager, and Sales Manager for Nordic Region and several Corporate Procurement Executives. In the Malaysian site we have talked to the General Manager, R&D Manager, Business Development Manager, Key Account Manager, MIS Section Head, NPI Manager, and Supply Chain Section Head. In Switzerland our informants were General Manager and Business Development Manager. The interviews have been recorded and transcribed and the content of the observations and the analysis has been cleared with the top management. The case has been written in the teaching case style. We use this narrative nature of the text to draw main observation regarding the interplay between vendor's process in outsourcing and vendor's competences.

EMS Industry

The EMS industry has its roots in the surge of outsourcing of manufacturing within the last thirty to forty years. Their clients are typically Original Equipment Manufacturers (OEMs), large manufacturers who traditionally have designed and manufactured products to sell under their own brand. The OEMs could come from various industries, such as communications (Sony Ericsson), computers (IBM), consumer electronics (Bang & Olufsen), automotive (Magna Donnelly), industrial (Danfoss), medical (GE Medical), commercial aviation (Boeing), defense (BAE) or other transportation industries (Modular Mining).

The main reason behind the rise of the EMS industry is the increasing needs of the OEMs to focus their resources on their marketing and design activities and their inability to cope in parallel with the increasing investments and management resources required by new manufacturing technologies. This led OEMs to outsource some of their manufacturing or even to divest some of their plants by selling them to EMS providers. Furthermore OEMs have come under intense pressure to reduce the costs of goods and some had to relocate their production to lower-cost countries. They were then faced with the choice of establishing their own production operation in an off shoring operation or to outsource to an EMS partner.

OEMs began to outsource the insertion of the electronic components on printed circuit boards (PCBs), in the process known as PCB assembly (PCBA), including it's testing some 30 years ago. This marked the beginning of the growth of Electronic Manufacturing Service providers (EMS), or Contract Electronic Manufacturers (CEM), who provide manufacturing services to OEMs. They manufacture products based on designs provided by OEMs, and sold by OEMs under the OEM brand name. Some of the leading tier-one EMS companies are Hon Hai Precision Industry (better known as "Foxconn"), Flextronics and Elcoteq.

Another type of company that participates in the EMS market are Original Design Manufacturers (ODMs), who provide design and manufacturing services to OEMs and brand them under their own name (Asustek, Quanta). ODMs are similar to the EMS providers, but they typically own the intellectual property rights (IPR) to the product itself, whilst the regular EMS providers build strictly to their customers' designs or sell their design services to their customers but in any case the IPR belong to the customer. EMS can make hundreds or even thousands of different products, whereas ODMs often specialize in only a handful of categories. Finally, the last group of companies in the EMS market are manufacturers who have upgraded their business from an EMS or/and ODM stage. They are known as Original Brand

Manufacturers (OBMs), whose characteristic is that they design, manufacture and sell products under their own brand name (*Acer*).

The top 10 EMS providers account for 70% of the sales of the entire EMS industry (Manufacturing Market Insider, 2007). EMS providers are segmented according to revenue, competence and global reach into tier-one, tier-two and tier-three EMS providers. Tier-one EMS providers are those organizations with sales figures over US$1 billion with a complete portfolio of competences and extensive global reach; tier-twos generate sales between US$1 billion and 100 million, have a degree of vertical integration and presence on more than one continent and may offer design services as well, but very few offer a combination of all three; and tier-three EMS providers are those with sales below US$100 million and are mainly smaller local companies specialized in, for example, PCB insertion for a particular market segment.

Large EMS providers predominantly serve computing and telecommunication industries. Companies like *Hon Hai*, *Flextronics*, and *Jabil* are looking for big deals worth several hundred-million or even billions of US$. These deals typically come from large OEMs like *HP, Apple* and *Cisco*.

In 2006, the EMS and ODM markets were worth more than US$ 200 billion. The EMS market still dominates the market with a share of 57% and is expected to be worth more than US$350 billion in 2009; more than US$440 billion in 2011. Despite these optimistic prognoses, EMS providers are margin-limited in their core manufacturing businesses. Within the field, mergers and acquisitions (M&As) represent one of the most obvious paths to growth. However, M&A activities seem to be saturated among large EMS providers, whilst there is still room for tier-two and tier-three companies to grow through mergers and acquisitions, also as a means of adding manufacturing or design competences, of extending geographic reach, or of capturing attractive customers or markets via the acquisitions. The number of cases in which OEMs are divesting their operations grew significantly in 2006, which broadened the number of possible acquisition targets. EMS market is still very attractive for new players, who increased in number in 2006.

The EMS market will continue to grow through 2010 but will certainly have to struggle with increasing competition from original design manufacturing (ODM) and original brand manufacturing (OBM) giants. According to Technology Forecasters, EMS revenue will grow 11% per year through to 2009, from $109.5 billion in 2004 to an expected $197.5 billion in 2009. Meanwhile, ODM revenue is predicted to grow 16% per year as sales rise from $69.5 billion to an expected $148 billion in 2009.

The Case Story of Escatec

In 1983 a young engineer sat up a fairly simple manufacturing facility in Northern Malaysia in order to produce power supplies for a multinational electronics device manufacturer. Escatec are still headquartered in Malaysia, and run by a diverse team of international executives. Over the years Escatec gradually built their competences by bringing manufacturing technologies like tool making, plastic molding and metal stamping in-house, thus enabling it to offer contract manufacturing of electronic and mechatronic (precision assemblies of electronics with engineering plastic and metal parts) products. Revenues and profits have grown steadily over the years to 176 mio UDS in 2006, providing the resources for organic growth as well as making possible acquisitions. E have built a stable portfolio of marquis OEM customers, mainly in industrial electronics, residential control, high-end consumer, and electrical products markets, while avoiding the known volatility in markets for computers and mobile phones. By 2002, Escatec were operating in four countries, with roughly three-quarters of their sales to European customers and one-quarter to North American customers.

In January 2003, Escatec made a major acquisition by buying a Swiss-based electronic manufacturer comprising advanced competences within electronic design, development and manufacturing. What may have seemed to be a counter-intuitive acquisition of a factory in a relatively high-cost location was in fact a highly strategic move on the part of Escatec to acquire advanced competences and to stake out a position in the heart of Europe; Escatec could then offer their European customers the convenience of local support in the same time zone, combined with low-cost Asian mass production.

With the efforts of both Asian and Swiss management teams, a new success formula seemed to have been born – a tier-two, vertically-integrated electronic design and manufacturing service provider with international reach that could offer a set of competences usually only found in tier-ones EMS providers.

Expansion of Escatec's Service Portfolio

Since its beginning, Escatec have been expanding their portfolio of services in the growing EMS market. The seed of all of the electronic manufacturing services, i.e., PCBA, was planted in Malaysia and during the early days required a large amount of manual work. The management's vision from the start was to make Escatec a one-stop EMS that could offer a full range of manufacturing services that are usually expected from tier-one EMS providers, but with the focus and flexibility of a medium-sized company. Each year Escatec have invested 3% of their turnover into new manufacturing technologies. Today, the company's electronics manufacturing is supported by recent technologies such as Chip On Board (COB), Flip Chip, Ball Grid Array (BGA), micro BGA and others that have enabled Escatec to assemble more technically advanced PCBs in large volumes and with exemplary quality.

Electronics manufacturing is quality-sensitive, and EMS providers must prove to their clients that they are able to manufacture at the highest quality levels, striving toward Zero Defects. Therefore, in building its manufacturing competences, Escatec have been paying extra attention to developing testing competences, both through designing and then contracting the fabrication of testing equipment locally and through developing a strong staff of qualified engineers. That has turned out to be a good move. Even better news was the realization that testing competences had grown to the level that could help expand their scope of services beyond post-manufacturing quality control. Escatec are now offering testing services to clients, either as a supplementary integrated service together with manufacturing, or as an independent service. Through acquisition of the Swiss company and subsequent know-how transfer from Switzerland to Malaysia, test competences have been greatly improved.

In the early days of PCB assembly, the management and engineers of Escatec started to ask themselves: "We can make a PCBA, but why not make the whole product for our customers? How could we do it? What would we need?" The answers were relatively straightforward. In order to manufacture and assemble a complete product for their customer, Escatec would need to develop new competences in mechatronic assembly, in some cases including complex coil windings and shielded higher level assemblies, ultimately putting together the whole product, a concept known in the industry as a box-build. In order to achieve this, Escatec needed to secure new sources in their supply chain for fabricated metal parts and plastic injection-molded parts, and also consider further backward integration to design and make the tooling such as injection moulds and progressive metal stamping dies. The only dilemma that remained was whether to subcontract these processes or to bring them in-house and to develop these competences internally. The company took the strategic decision to go for the latter and try to capture as much of the value chain as possible, and hence be in a position to manage and control quality and lead-times.

Escatec backward integrated into plastic injection and metal stamping, including the design and production of tools and dies, in Singapore in 1987. Gradually, following cost pressure and customer demand, Escatec moved metal stamping, coil winding and plastics molding to Malaysia and Indonesia, and tool design and tool-making to Shanghai. The scene was set for a new competence: the box build, which quickly was realized successfully with challenging products such as a coffee machine for the high-end home market in the US running in very large quantities and requiring over 50 tools for plastic and metal parts – a product that Escatec never could have been considered for before. The final ingredient, in the form of global supply chain management, was added and Escatec could then fairly claim that the goal of being a one-stop tier-two EMS had been achieved.

From around 2000 it became clear that OEMs have been starting to send signals to the market that they were getting ready to consider the next level of outsourcing, i.e., outsourcing a process that had previously been considered too close to OEMs proprietary or core competences to be consider for outsourcing: design and development (D&D) of PCBAs and complete products. This was a paradigm shift that opened the doors for new customers, new projects, and even new strategic goals, like becoming an ODM, which now had to be considered as a strategic option for Escatec.

The move to acquire the Swiss company in 2003 was a carefully planned strategic move: a step to bolster the company against rising competition within the EMS market, by augmenting its technology portfolio and adding an important geographic location. Before the acquisition, Escatec only had access to very few resources within D&D. After the Swiss company became part of the family, deeper D&D competences were integrated into the value chain and 36 D&D people began working on various projects. Today, customers can approach Escatec with D&D tasks and they can even choose another manufacturer once the design job is done. But the vast majority of customers (98%) stay when it comes to transposing product plans into manufacturing processes. The new profile of competences also brought the company closer to its strategic goal of being an engineering service provider with special D&D skills within areas such as PCB design, high tech multi-layer boards, and even micro-opto-electro-mechanical systems (MOEMS), a new field moving from the macro toward the micro and even nano world. “Some customers arrive with a preliminary prototype or even just a concept of a product, and we can offer them design for manufacturing. We can do electronic and mechanical design, as well as PCB layout according to their functional specification. We can take a product through prototyping or manufacturing and can offer them significant cost reduction versus their in-house activities," says the D&D director.

The integration of various operations is carefully balanced against the regional deployment of plants and functions. Constant adaptation to changing competitive conditions seems to be the key to Escatec's strategy. The founder and president notes: "One of our oldest clients came from a company in Singapore that we acquired in the early 1990s. This production, which employed about 1.000 people, used to be situated in Singapore. After some time, it became too expensive so we simply started to transfer production over the Causeway from Singapore to Johor Bahru, Malaysia, where labor costs were half. Later on, when mass production reached levels as high as 8 million units a year, it even became too expensive to produce Malaysia, and so in 1999 it was shifted to Indonesia and again the labor cost was halved."

The integration of D&D competences to Escatec's industrial panoply has created a business activity with high value creation and high margins, which connect customers more intimately to the company as human relations are formed along with a deepened collaboration process. The D&D competence, whilst it must be a standalone profit centre, is a means of securing projects in an early stage then being in an incumbent position to capture the tooling and manufacturing business. The director of Business Development explains: "We can secure customers at any stage in our value chain. Customers can enter and exit at any step, but generally, once they're in, they stay in."

Having an impressive number of services delivered throughout global locations is one thing, but integrating this into an efficient and reliable process is another. Nevertheless, Escatec have confronted the challenge.

Integration of processes

And indeed, Escatec coined a specific name for their service provision process: E^2MS (Electronic Engineering Manufacturing Service), thus emphasizing the synergies gained from expanding and integrating their service portfolio (See Figure 1).

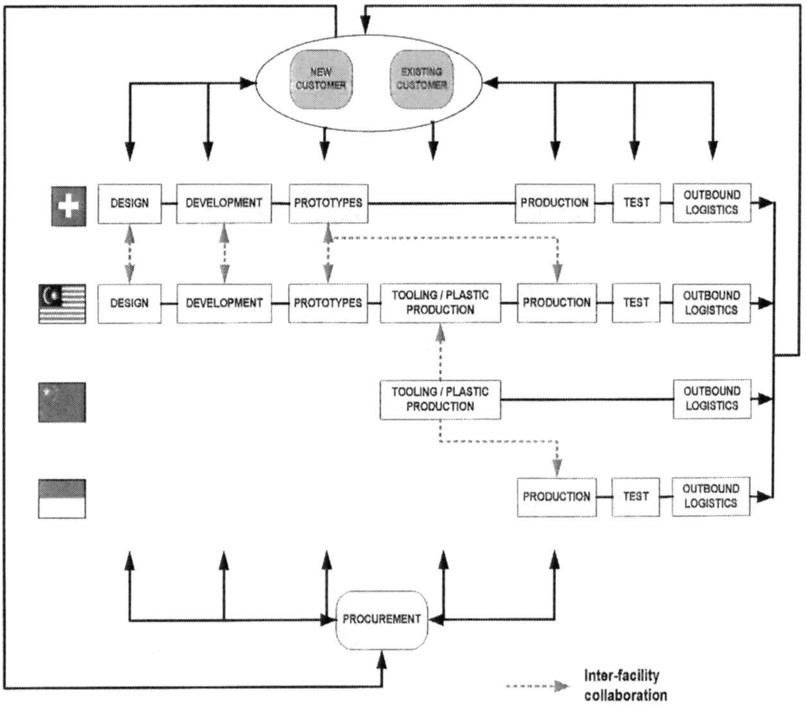

Figure 1. E^2MS process

It starts with the design phase and continues with the development, prototyping, tooling and production phases and continues all the way to the testing phase. These phases are supported by procurement and supply chain functions, which are being conducted in and across the different locations of Escatec's facilities in Malaysia, Switzerland, Indonesia and China. The fine filament is constantly being adapted in order to satisfy the alpha and omega of the EMS market: its customers—new ones as well as existing.

But in the highly competitive tier-two EMS market, winning the outsourcing contract is not that easy. So E has developed a systematic way of approaching them. Escatec's sales people are located in their most important markets: Switzerland, Germany, the UK, the Nordic countries, and the US, and their main tasks are to identify potential customers and to secure contacts. It is important at this stage that both the customer and Escatec become acquainted with each other, so that when the customer writes a request for quotations (RFQ), Escatec will have substantial knowledge about the customer and hence be in a better position to win the contract.

The integration of services strategy is paying off in this process. A customer can be captured anywhere across the value chain. The sales teams are backed by the business development and D&D units, and operate according to various funnel models, addressing each market niche individually, both with respect to geography and the products. Potential customers are always informed about the design, development, tooling and manufacturing competences of the Escatec Group. "Sometimes Escatec have customers that are only interested in D&D in Switzerland. However, they may have the need for manufacturing more than 100,000 units per year, which makes Malaysia an interesting location", explains the Business Development Manager for Escatec Switzerland. The Chief Operating Officer emphasizes that the various alternative manufacturing locations are one of Escatec's competitive advantages over other tier-two EMS competitors. The project team evaluates a product based on its complexity and the customer's requirements, and then selects between Switzerland, Malaysia or Indonesia – or even more powerfully, starting a project in the D&D and new product introduction (NPI) phase, then managing a seamless internal transfer process for the customer when the product is ripe for mass production at Asian costs.

When the contract is signed, everything must then occur very quickly. Speed is crucial, as the customer often wants to see the product the next day. That's where the NPI team steps in. Escatec's approach to the NPI project management concept was formalized at Escatec Switzerland, and then rolled out across the entire Escatec Group. The main mission of NPI is literally to "introduce" a new product into manufacturing, taking it through the often iterative process of design, prototyping, and small series production until proven and stable. Depending on the point of departure in the E^2MS process, there could be three types of NPI projects:

i. The first type is a *pre-project*. It includes only development and design activities for the client.
ii. The second type is a *development project* that starts with the development, goes through the design phase and ends with product prototypes being developed for the client.
iii. The third type of project is the *industrialization and displacement project* of which the main aim is to prepare the assembly and/or manufacturing of the box-build for a product that has been pre-designed outside of Escatec. This type of project starts at the tooling stage of the process.

The duration of these three projects can vary. "The industrialization and the transfer of manufacturing (displacement) project can't be conducted in less than one month. But if we start the NPI from scratch, which eventually involves all stages from development to production, it can last for more than one year", adds the NPI Manager in charge of introducing the NPI process in the Malaysian facilities.

Providing services on the extremely cost sensitive EMS market is an intense and often adrenalin-raising process. The NPI team, which consists of experts (often from both the Swiss and Malaysian sites) from various departments within Escatec, has to secure a smooth transfer of the product into the production phase. The team must also ensure that the project is being conducted simultaneously across all phases of E^2MS in the reality of geographical distances, which all in all is not an easy task.

Since the acquisition of the Swiss company, Escatec have worked to bring up the level of their D&D centre in Malaysia. A veteran member of Escatec's design team in Switzerland for over ten years, the D&D Manager is on a long-term assignment from Switzerland to further build a D&D team in Malaysia. Upon his arrival he was faced with many challenges and was certainly not spared from culture shock: "The approach towards Design and Development and human resources management are completely different in Malaysia and Switzerland. Here I had to learn how to recruit top engineers, how to train and coach them, how to monitor their work, motivate and retain them in the face of big name MNCs all around us".

In the cost-competitive contract manufacturing market, EMS providers often compete to purchase materials under competitive conditions, which bring the procurement function into specific focus. In order to win a contract, the most important thing is that procurement has competitive prices. Around 70%, for some products even 85%, of the production costs are allocated to material costs. Escatec can only control the remaining 15-30% of production costs when bidding. That is why the NPI team and the Procurement function must work with their suppliers with the same fervor that they use to collaborate with customers. "The trick is how much are we able to pool components. In order to promote the common material we need to have an Escatec Approved Vendor List (EAVL). We have our own preferred manufacturers and we are trying to convince our customers to use this preferred list. The customers are therefore able to create the Bill of Material (BOM) based on our EAVL". By using specialized quotation software, E have completely automated and optimized the process of transforming the customer's BOM into a request for quotes (RFQs) disseminated to their preferred suppliers via a special web portal allowing E to receive quotes back in as little as just one day, or in up to two weeks for more complex products.

Another complex issue in the E^2MS and NPI processes is how to successfully manage the production of plastic parts and tools.

These operations are not perceived as the company's "money makers", rather, they're strategic competences meant to bring in new customers for complete box-build services, and could eventually secure the entrance to new geographical markets, like the Shanghai-based tooling facility. However, this strategy is not always easy to support in the trenches or the operation. For the Divisional Manager in charge of production of plastics, it can be a real headache to be in the position of constantly having to run small lot sizes of hundreds of different parts required by the just-in-time (JIT) demands of box-build assembly. "I hope that they know how difficult it is to regularly re-layout a plant when it entails moving such heavy machines around and to set up moulds on machines for short production runs".

To gain a new customer appears to be a very complex challenge. But retaining an existing customer is no less problematic. However, E successfully manages to retain their long-lasting customers' loyalty from whom their new product pipeline predominantly comes from. An experienced electronics design and manufacturing veteran who Escatec's Switzerland plant, posits that "The road to success is based on total customer satisfaction and deep customer relationship, starting with engineering services moving through specialized electronic manufacturing right through to after sale service and support". The President frequently calls or visits all of his customers from the largest to the smallest. His approach builds trust and Escatec's fast and reliable response to fixing problems makes his customers ready to rely on his organization. The latter is the responsibility of key account managers. At Escatec they are the customer's ambassadors, carefully monitoring the provision of service, from design to manufacturing; from procurement to delivery. "We have multi-layered relationships. This is because we are a customer-focused organization".

Conclusion

We have presented a case of a tier-two EMS service provider who has been adding various skills, knowledge and technologies into a set of competences that have been integrated in the meaningful process. The vendor's process is cyclic in its nature, with equally strong emphasis on getting the new and retaining the existing customers.

The whole set of competences on the corporate level has emerged as a result of the strategies that led towards the geographic expansion and the expansion of the service portfolio. "We will evaluate the product based on its complexity, customer's requirements and then select Switzerland, Malaysia or Indonesia", is how the Chief Operating Officer (COO) explained how Escatec manage alternative locations for manufacturing. Escatec's value chain has been built around the process that consists of the design, development, prototyping, tooling and plastics production, electronic and mechatronic manufacturing, testing and outbound logistics. Forward and backward integration of services has been initiated by developments in technologies that are characteristic for the EMS industry. Various teams within E have contributed to this. Their breadth and depth enable Escatec to be able to introduce new technologies. "We are fast in investigating, learning and implementing new technology. Other EMS' tend to stick to their technologies", claims the COO.

Modular structure of the process enables Escatec to capture a customer anywhere across the value chain. Vertical integration combined with a relationship management helps Escatec to maintain very high customer retention rate. Five major customers have been with Escatec for more than 20 years in average contributing to Escatec's turnover with 47%. The main reasons why the existing customers are choosing to entrust new projects to Escatec are corresponding to the competences embedded within Escatec's process, organisational units and on the corporate levels.

The paper has been written in a teaching case style. Its academic relevance can be applied in teaching graduate, post-graduate and executive classes in international operations and outsourcing. Further on, the case can be an exemplifier in studying organizational processes and competences.

From the industry perspective, the case of Escatec is a show room of how a tier-two EMS company can have the structure and competences of a tier-one EMS giant. The road to a successful one-stop-shop type of EMS operations lies in the formula:

Establish rich portfolio of competences - Integrate the competences into a process – Ensure that process depicts the nature of the industry – Develop a modular process that will enable capturing of customers anywhere along the value chain – Continuously improve competences that mostly contribute to organization's performance

References

Blois K., (2004). "Analyzing exchanges through the use of value equations", *Journal of Business and Industrial Marketing*, 19, 4, pp. 250-257

Click R.L., Duening T.N. (2005). *Business Process Outsourcing: The Competitive Advantage*, John Wiley & Sons, Hoboken

Cullen S., Willcocks L., (2003). *Intelligent IT Outsourcing: Eight Building Blocks to Success*, Butterworth-Heinemann, Oxford

Feeny D., Lacity M., Willcocks L., (2005). "Taking the measure of outsourcing providers", *MIT Sloan Management Review*, 46, 3, pp. 41-48

Jenster P.V., Pedersen H.S., Plackett P., Hussey D., (2005). *Outsourcing Insourcing*, John Wile & Sons, Chichester

Kern T., Willcocks L., van Heck E., (2002). "The Winner's Curse in IT Outsourcing: Strategies for Avoiding Relational Trauma", *California Management Review*, 44, 2, pp. 47-69

Levina N., Ross J.W., (2003). "From the Vendor's Perspective: Exploring the Value Proposition in Information Technology Outsourcing", *MIS Quarterly*, 27, 3, pp. 331-364

Liang H., Xue Y., (2004). "Coping with ERP-related contextual issues in SMEs: a vendor's perspective", *Strategic Information Systems*, 13, pp. 399-415

Manufacturing Market Insider, 2007

Mohr J., Spekman R., (1994). "Characteristics of partnership success: Partnership attributes, communication behaviour, and conflict resolution techniques", *Strategic Management Journal*, 15, 2, pp. 135-152

Oza N.V., Hall T., Rainer A., Grey S., (2006). "Trust in software outsourcing relationships: An empirical investigation of Indian software companies", *Information and Software* Technology, 48, 5, pp. 345-354

Pinnington A., Woolcock P., (1997). "The Role of Vendor Companies in IS/IT Outsourcing", *International Journal of Information Management*, 17, 3, pp. 199-210

Rochester J., Rochester, H., (1995). "Advantages and Disadvantages of Outsourcing", *Proceedings of the International IEEE conference*, pp. 77-82

Stjernström S., Bengtsson L., (2004). "Supplier perspective on business relationships: experiences from six small suppliers", *Journal of Purchasing and Supply Management*, 10, pp. 137-146

Vollman T., Condor C., (2000). *Outsourcing Project Workshop Presentation* quoted in Jenster at al. (2005)

Willcocks L., Choi C.J., (1995). "Co-operative Partnership and 'Total' IT Outsourcing: From Contractual Obligation to Strategic Alliance?", *European Management Journal*, 13, 1, pp. 67-78

Zviran M., Ahituv N., Armoni A., (2001). "Building outsourcing relationships across the global community: the UPS-Motorola experience", *Journal of Strategic Information Systems*, 10, pp. 313-333

SUPPORT SERVICES IN DEVELOPING TECHNOLOGY B-TO-B RELATIONSHIPS

JUHA-PEKKA KOISTINEN
Co-founder, Globate Ltd.
Runeberginkatu 4 AC 49, 00100 Helsinki, Finland
juha.pekka.koistinen@gmail.com

PEKKA ESKOLA
Department of Industrial Engineering and Management, Helsinki University of Technology,
PL 1000, 02015 TKK, Finland
pekka.eskola@utu.fi

Abstract

Especially companies operating in business-to-business markets have acknowledged the benefits that can be gained from long customer relationship. As buyers operating in b-to-b markets make the purchases in order to achieve organizational objectives, they are generally interested in the value that can be generated with the equipment rather than only its features. The research problem of the study was: how can support services be used in developing high-technology b-to-b relationships?

The structure of a b-to-b relationship was discussed using the interaction model consisting of interaction, interacting parties, environment, and atmosphere. The authors argued that the development of a relationship could be described best as seven stages framework. Koistinen & Eskola (2008) development stage model includes initial stage, establishment of relationship, exploration stage, expansion stage, stable stage, disengagement stage, and consequences management.

By using support services it is possible to increase customer value, thus leading to customer satisfaction, which in part leads to customer commitment. The theoretical framework was empirically tested by applying it to high-technology b-to-b marketing context. The empirical research was conducted as a qualitative research by interviewing six representatives of partners of the case company. The case company in this study was Nordic ID, a Finnish supplier of handheld devices used in wireless automated data capture.

The results of the study indicate that systematic relationship development is possible. However, the systematic aspect appears to be heavily grounded on the factors through which relationships develop over time in the first place: social exchange between individuals, information exchange, mutual trust, and long term mutual commitment.

On the basis of the research, support services can effectively be used in establishing new customer relationships. Their importance for the customer supercedes even the aspects associated with the product itself. When the relationship is already established, the effect of support services changes. Rather than actively developing it, support services keep the relationship active and open for more interaction. It is at this point when information and social exchange together with trust become especially important, as they ultimately develop the relationship.

According to the results of the study, there can be suggested many managerial implications. The authors argue, firstly, that managers cannot overestimate the meaning of information sharing in support services. Secondly, managers should consider the possibilities to establish an open information sharing within a network of interconnected companies. This network could be built from companies that complement each other's abilities and advantages. Thirdly, development associated with each company's technological core competence could be coordinated by partners possessing necessary knowledge of the subject. This would increase innovativeness and efficiency of research and development. Finally, some responsibility of customization could be moved to the partner possessing most knowledge about the subject, thus lightening the workload of the producer of the core product. Over time the companies in the network could also be able to provide support for the products of the other companies when possible.

Keywords: support services, business-to-business relationship, high-technology, relationship development

BACKGROUND FOR THE STUDY

Marketing strategy based on long-term customer relationships is nowadays a concept followed by many companies operating in b-to-b markets, but also by those concentrating on consumer business. However, this has only been the case for a relatively short time. Even though close relationships with customers have historically been important for merchants, the theories associated especially with consumer marketing have been emphasizing the significance of temporary, product-related factors associated with the seller's offering in winning deals, even when the customers are already familiar to the seller through previous business.

The role of support services is becoming increasingly important in b-to-b marketing. Despite new industrial products and equipment being more reliable than their predecessors, they are becoming more complex and advanced than before (Markeset & Kumar 2005, 54). Besides enabling the effective use of equipment, support services are also recognised to be a powerful mean of adding value to the offering as well as being a differentiation tool, and thus a transition from pure manufacturing company to service provision on at least some level is often found to be a direction for development.

This study concentrates on b-to-b relationships and customer support services offered in b-to-b markets. The aspect of relationship receiving most interest in this study is the development of relationships. The research problem of this study is: *how can support services be used in developing high-technology b-to-b relationships?*

The research problem will be approached by first discussing the structure of relationship and processes associated with relationship development. After this the elements of support services are described and role discussed. The empirical part of research is concerned with providing and analysing connections between different support services and relationship processes in the selected case company. Finally, at the end of the study, there are suggested some managerial implications.

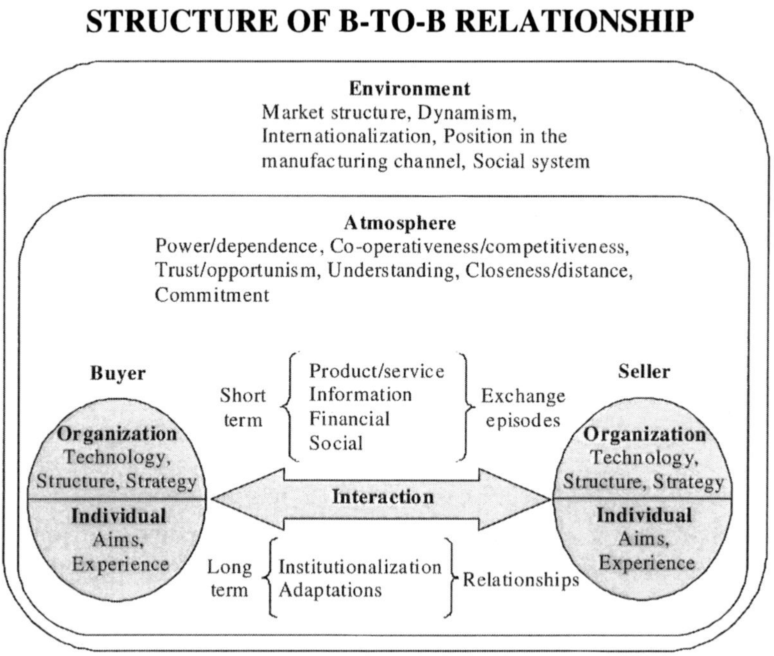

Figure 1. An illustration of the interaction model (modified from Håkansson 1982, 24 and Hallén & Sandström 1991, 118-122).

In this study interaction model developed by IMP Group provides the framework with which structure of a b-to-b relationship will be discussed (see Figure 1).

An above model was chosen because of its built-in suitability for studying b-to-b relationships, as the model was developed by researching specifically relationships in b-to-b markets. Interaction and communication between the two companies is an integral part of their relationship. *Interaction* includes both short-term and long-term elements. Actions such as individual phone calls or exchanges associated with products, information, money, or social aspects represent the short-term aspect. Interrelated actions form episodes and similarly interrelated episodes form sequences. Relationships are built of sequences and represent the long-term nature of interaction. (Holmlund 1996, 49-51; Håkansson 1982, 15-17.) Over time the companies become institutionalized to common ways of working and also make adaptations that bring the companies closer to each other.

Interaction cannot exist without interacting *parties*. In b-to-b relationships the parties are defined by both organizational and individual factors, and thus characteristics of both aspects affect the relationship. Organizational factors include technologies used by the companies, structures of the organizations, and strategies pursued by the companies (Gross et al. 1993, 4; Ford 1990, 47). Aims, experiences, backgrounds, and characteristics of the individuals operating in the companies considerably affect the actions of the companies and the development of the relationship. Personal relationships between the contact persons in the companies are an important part of the relationship's infrastructure, as well.

A relationship between companies takes place in a certain *environment*. The environment affects the companies' decisions and activities. The effect of environment can be examined by dividing it into effects incurred by elements of the environment. The environment can be divided for example into political-legal environment, economic environment, social-cultural environment, technological environment, and natural environment. (Hutt & Speh 1998, 76-78; Dwyer & Tanner 1999, 108.) However, Håkansson (1982, 20-21) argue that when discussing b-to-b relationships, aspects associated with the market where the companies are operating becomes especially important.

The fourth of the main elements is relationship *atmosphere*. It can be described as being an emotional setting in which the relationship takes place. Relationship atmosphere affects all of the other elements of the relationship and is also affected by them. (Hallén & Sandström 1991, 112-114.) However, Sutton-Brady (2000, 2) adds that as relationships evolve, atmosphere should be primarily seen as the product of the interaction.

DEVELOPING RELATIONSHIPS

The development of a relationship can be described by using different stages or levels of a relationship. For instance, Ford (1990, 45) sees the development of relationship in five stages,

Dwyer, Schurr and Oh (1987, 15-20) also describe five stages and Halinen and Tähtinen (2002, 173) concentrate on the dissolution process in more detail.

Koistinen & Eskola (2008) have made an extensive literature review of different development stage frameworks of b-to-b relationships. They suggest that relationship development stages can be illustrated best as a six-step-model (see Figure 2).

Since the research problem of this study is "How can support services be used in developing high-technology b-to-b relationships?", the last three stages of Koistinen & Eskola (2008) development stage model will be now studied in more detail: expansion stage, stable stage, and disengagement stage.

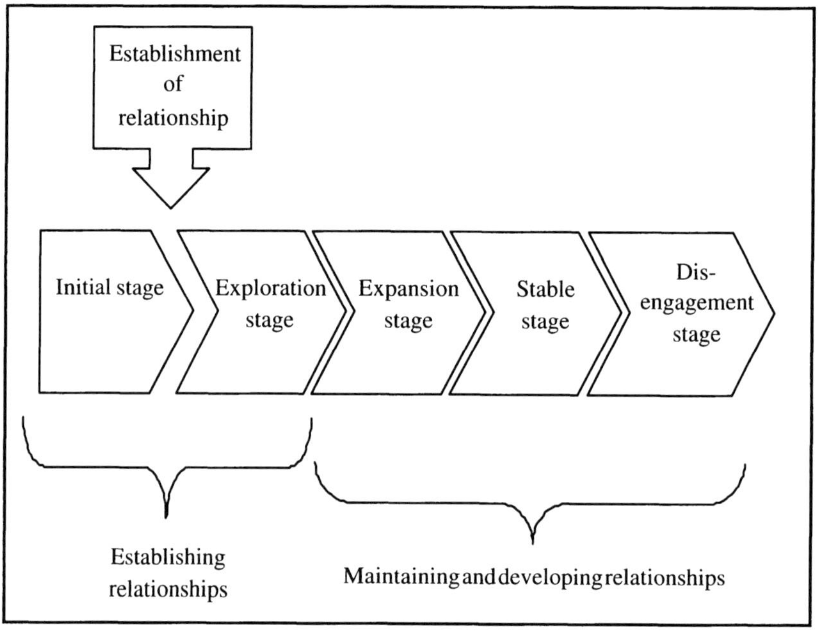

Figure 2. Koistinen & Eskola (2008) development stage model

Expansion stage

In expansion stage companies' interaction and their interdependence increases. This results in more experience about the other party and it reduces the distances and uncertainties in the relationship associated with technology and cultural differences, for instance. Halinen (1994, 280) states that increasing experience and interaction between the companies also contributes to the trust between individuals in the companies, as trust is built through past common experiences. Trust between the parties is essential in order for the relationship to develop, since efforts made to develop the relationship expose the companies to the actions of the other party and make possible the abuse of the situation. According to Coulter and Coulter (2002, 37) trust can be defined as being built on four factors associated with the interacting parties: confidentiality, honesty, integrity and high ethical standards.

Trust can be created and uncertainties associated with distances can be reduced by mutual adaptations made by both companies. Motivations for adaptation include cost reduction for both companies as well as improved over either the supply or buyer markets in which the companies operate. Besides facilitating co-operation, adaptations made by the companies also signal their commitment to the development of the relationship.

Stable stage

Stable stage is considered to be the highest possible development stages of a buyer-seller relationship, and as such an aim which relationships are striving to achieve. At this stage the parties have achieved a high level of satisfaction with the relationship and this makes it difficult for competing companies and their offerings to intervene, even if they can offer similar or even superior levels of benefits. (Dwyer 1987, 19.) The companies' mutual importance to each other and continuity of the relationship mark the stable stage. Ford (1990, 50) argues that these are achieved after large-scale deliveries of continuously purchased products or several major capital purchases. Mutual trust makes it possible for the relationship to withstand possible short-term problems.

In stable stage the companies have extensive amount of experience about the other party and this experience creates trust and standard operating procedures which reduce distances and uncertainties to a very low level. For example uncertainties related to social distance are reduced to a very low level. Even though,

Ford (1990, 50-51) argues that there are three factors resulting from the added experience needing special attention.

First of the factors is associated with contact patterns. When the companies are in stable stage, they are bound to have developed extensive contact patterns. These contact patterns may become so complicated, that communication between different departments may become uncontrolled, thus disturbing effective operation and even causing conflicts between departments. The second point to take into consideration is the strong personal relationship developed between the individuals from two interacting parties. The relationships can cause difficulties when organizational and personal interests collide. The third point about low level of social distance to consider when in stable stage is the level of personification of a company in one individual representative. This personification may be of great value and sometimes even necessary in order to acquire business, but creates problems if this individual must be replaced for some reason.

Disengagement stage

Disengagement stage begins the dissolution of the relationship. In disengagement stage functional company level connections between the companies are cut but it is difficult to determine when a business relationship ceases to exist. Possible reasons for dissolution of a relationship include assessments of relationship potential, performance of the other party and outsiders' actions (Ford 1990, 50). The ability and desirability for a company to end a relationship depends on the attractiveness of available alternatives to the current relationship. Zeithaml et al. (1996, 33-34) points out that the disengagement of a relationship incurs costs especially to the supplying company as new customers need to be acquired.

SUPPORT SERVICES AS A RELATIONSHIP MANAGEMENT TOOL

Customers need assistance in order to gain maximum value from any new high-technology product. Support services are services that are provided by the supplier in conjunction with the bought equipment in order to help the customer gain value from it, but also to differentiate the supplier's offering from those of competitors. Markeset and Kumar (2003, 381) state that the customer's need for support depends on customer's knowledge and capabilities, characteristics of the product, and the operating environment of the equipment. In general at least some support services are needed with all equipments. Warranty, installation, maintenance and repair, user training, and technical support are support services for example. Support services can be divided into those supporting primarily the product and those supporting primarily the customer, and also into pre-purchase and post-purchase support services, depending on whether they are of use before or after the purchase (Mathieu 2001a, 40; Oliva & Kallenberg 2003, 168).

Support services affect development of the customer relationship both by the route of continual evaluation and also by specific critical episodes. B-to-b customers are generally interested in the total value produced by the equipment rather than merely the technical characteristics of it. As value perceived by the customer is important in winning deals and consequently first step towards a long-term customer relationship, supplying companies try to create as much value for the customer as possible (Heskett et al. 1994, 166; Ravald & Grönroos 1996, 19). Value can be added by increasing the benefits perceived by the customer or reducing the sacrifices and costs incurring to him. Support services can be used in both cases but their role is bigger in increasing the benefits. In b-to-b marketing the service aspect of the offering is integral, as good services provided with equipment are in many cases considered more important by the buyer than features of the product itself (Burger & Cann 1995, 91; Cann 1998, 393-394). Besides the value of the individual exchange episode, value has also a relationship aspect; a long-term customer relationship itself brings additional value. The value of the relationship itself becomes evident in feeling of security and minimized purchasing risks, for instance.

Maximizing the value perceived by the customer is important also for reasons other than winning the first deal. Customer value can be increased by support services, thus creating customer satisfaction that contributes to the development of a long-term relationship. Patterson and Spreng (1997, 416-418) state that after the first purchase the related satisfaction or dissatisfaction becomes the most important factor determining the development of the relationship. Besides through value, support services are also able to increase customer satisfaction directly. High customer satisfaction is important as it drives customer commitment, which describes the companies' willingness to maintain and develop the relationship (Lam et al. 2004, 308; Ravald & Grönroos 1996, 19). Customer commitment includes for example re-purchase behaviour and spreading positive word-of-mouth about the supplier, both of which are factors increasing the sales of the supplying company.

Besides affecting the elements of the chain from customer value to customer commitment and thus the continual evaluation, support services also affect relationship development directly by critical episodes. A critical episode has positive, neutral, or negative effect on the development of the relationship. Critical episodes can be directly associated with the support services. For example, service of equipment can become a critical episode with negative effect for the customer's business if it is not done well. (Schurr 2004, 6-7.) Similarly, especially well functioning technical support can contribute in an episode becoming critical with positive effect.

RESEARCH DESIGN

The theoretical framework of the study is now empirically tested by applying it to a high-technology b-to-b marketing context. The selection of a case company was especially influenced by two factors. Firstly, it was assumed that in high technology companies the role of support services would be bigger because of the complexity of the products involved. Secondly, it was also assumed that the concept of long term business relationship would be more relevant in b-to-b than in b-to-c context.

The case company in this study is Nordic ID, a Finnish company with headquarters in Salo, Finland, where it was also founded in 1986. The turnover of Nordic ID is about 11 million euro and the company has personnel of 45 people. The business of Nordic ID consists of designing and selling wireless handheld data collection terminals. These terminals can be used for various purposes, depending on the special needs of the customer's industry and functional characteristics. The main idea of wireless handheld terminals is that when an action concerning an article is done, the identification information of the article (for example, in the form of a barcode or a RFID tag) is read with the terminal. The terminal communicates with the company's information system, and the information about the article is immediately available to the whole organization. This makes it possible to control for example very efficiently the inventory levels, or to follow the status of a delivery.

The empirical part of the study is conducted as a qualitative research by interviewing six representatives of partners of the case company. The persons to be interviewed were suggested by the case company. This can be seen either as a good or a bad thing in relation to the quality of data collected. In this study, however, since it was also in the case company's interest to gain useful results, this can be seen as a positive factor. By having the persons to be interviewed suggested by the case company it was ensured that appropriate people were chosen for the interview. In total eight interviews were conducted and in two of these the interviewee was the contact person inside Nordic ID. These two interviews were conducted to gain background information about the area of business in question and also about current practices inside the case company. The other six interviews involved representatives of partner companies of Nordic ID. The number of interviews conducted was relatively small, but this was not considered as a major limiting factor since the persons to be interviewed represented the key knowledge of the subject in the context of the case company and its business environment.

In order to understand why the six interviewees were chosen from partner companies of Nordic ID, it is essential to recognize the role of the partner companies in Nordic ID's business. The selling is done by both Nordic ID itself and associated partners. When an order is placed, it goes through partner to Nordic ID. The delivery to end user is organized by Nordic ID, but the invoice for the transaction is sent by the partner, who also takes care of the installation. Nevertheless, direct connection with end users is used in large projects. In some relationships that have existed for a long time, direct connection with the customer is still used instead of integrators.

Figure 3 illustrates the organization of after-sales service by Nordic ID. As with transactions, partners have an important role also in after-sales service.

When a problem occurs, it can be reported directly to Nordic ID, or the partner can mediate it. Problems are normally reported via telephone or E-Mail. In the next step, the partner is informed about the response to the problem. Equipment needed to fix the problem can be sent directly to end user, or they can be mediated by the partner. The invoice for the support is sent to the partner by Nordic ID.

All six partner companies represented by the interviewees are important partners of Nordic ID. Pro Component Oy, Vilant Systems Oy, and Aldata Solution Finland Oy are all located in Finland and can be described as system integrators that use the products of Nordic ID in their solutions. Pro Component Oy operates in b-to-b commerce of solutions including servers, workstations, point-of-sale devices and data collection devices. Vilant Systems Oy specializes in RFID (Radio Frequency IDentification) technology and supplies logistics solutions for industry and retail industry. An important part of these solutions is software produced in-house? Aldata Solution provides solutions associated with supply chain management, point-of-sale, warehouse management and logistics to retail industry. They use Nordic ID's handheld devices as part of their solutions.

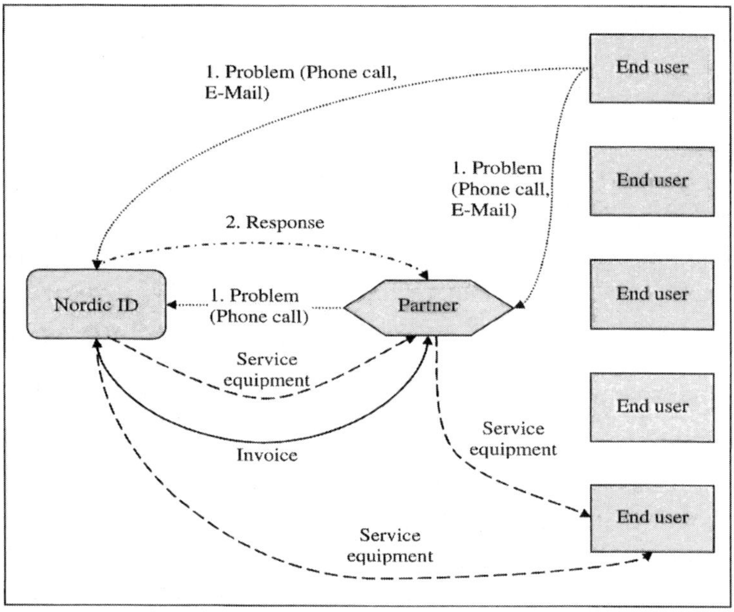

Figure 3. Organization of after-sales service by Nordic ID

Spar UK Ltd., Woolworths Group plc, and Midlands Co-operative Society Ltd. are UK-based retailing companies that use Nordic ID's products in their business operations. This brings new perspective to the study in two ways. First, international aspect is achieved by including the UK-based companies and English could be used as communication language. Second, the group of three retail industry companies that can be considered as end-customers balances the set of three system integrators that use the equipment as part of their solutions. Spar UK Ltd. is part of international Spar that has nearly 16000 stores in 34 countries. Woolworths Group plc. is a retailing chain that has 823 stores across the UK? Finally, Midlands Co-operative Society is an independent retail society operating in the UK. It has approximately 375 trading outlets.

RESULTS OF THE EMPIRICAL STUDY

Support services in developing existing relationships

Whereas all the interviewees agreed that business relationships evolve, the possibility of being able to systematically develop them divided opinions. Four of the interviewees think that systematic development is possible. It is worth of noticing that all the interviewees in the integrator group believe in systematic development. This illustrates their relationship-oriented approach to marketing; relationships are seen as providing future opportunities. Several means to achieve systematic development were brought up by the interviewees. These are generally associated with social exchange, information exchange, mutual trust, and long term mutual commitment. The social exchange between the companies generally makes information exchange between them more flexible and effective. When this is combined to open and frequent information exchange and communication, systematic relationship development can be achieved.

When information exchange is rich in content, it affects the actions of the interacting companies and helps them become better both in their own special business areas and also as a team. According to an interviewee, increasing trust is an essential mean of systematic relationship development. In addition, mutual long-term commitment together with long-term strategy and mutual motivation were also brought up by a respondent as constituting the process of systematic relationship development. One interviewee stated that systematic development of a relationship is not possible since flexibility and understanding are needed instead of a rigid system if the relationship is to develop. Even though, flexibility and understanding are strongly associated with adaptations and high-content information exchange that are considered as important elements of systematic relationship development by the other interviewees. This highlights the importance of those elements in the relationship development, whether this development is considered systematic or not.

Three of the respondents consider their relationship with Nordic ID as being systematically developed. The development of the relationship with Nordic ID seems to be strongly tied to information exchange and based on problem solving. Information coming from the initiative of Nordic ID includes for example product information and in some cases customer leads for the integrators to follow. This way of interaction functioning relatively heavy on the basis of problem-solving is found effective and natural by the interviewees.

When asked about factors that influence their willingness to continue a business relationship with a partner, several factors were brought up by the interviewees. Trust appears to be one of the most important of these factors. This is not surprising since developing trust is an essential process in relationship development. In this context trust seems to have three dimensions. Firstly, being able to trust that the partner is committed to the relationship and motivated to its development in the long run appears to be highly important because of time and resources put into the relationship. This aspect of trust can be gained only by building mutual commitment over time and signalling it by making adaptations for example. The second aspect of trust is that one has to be able to rely on the partner to provide support when needed even when it is not profitable on the short run, and also to comply with everything that has been agreed on; one must be able to rely on the information given and agreements made with the partner on areas such as schedules and prices.

> *There has to be trust, things that have been agreed have to be done. They are important, trust and the right nature of partnership. That you stand by your partner. Partners are chosen and then you work with them.*

Third aspect of trust in this context is that one has to be able to trust the partner to be efficient in his side of business. As stated by two interviewees, the trust associated with the partner's own business can be increased by helping him. This is done by offering help from another viewpoint, for example, by providing information that has been found out when interacting with one's own customers. As stated before, open communication together with information and social exchange are important factors in building trust. One interviewee described general responsiveness in the case of an occurring issue as the single most important aspect of communication whereas especially effective co-operation together with problem-solving abilities were found to be important factors affecting willingness to continue the relationship by several respondents.

> *I would highlight the importance of communication and open co-operation and problem solving, because if they don't work, nothing works. If price is the problem you talk about it and you solve the problem, if technology is the problem you can talk about it and usually it can be solved.*

Besides organization-level elements also inter-personal relationships are a factor affecting the companies' willingness to continue co-operation. The definitive role of individual persons operating in the companies was brought up as having a two-dimensional role in this context. Firstly, according to an interviewee, personal views and experiences greatly affect willingness to operate with a given partner. Secondly, strong inter-personal relationships are seen as very important building blocks of lasting co-operation, and thus developing these relationships between individuals is brought up as an effective mean to develop the business relationship itself.

Technology used in the solutions is also factor contributing to the willingness to stay in a relationship. The technical characteristics, capabilities, and the level of expertise of the companies for example affect the amount of adaptations needed, amount of trust needed, and also how the interaction between the companies needs to be organized. New solutions and technology were mentioned by four respondents as an important prerequisite for a lasting partnership. The price factor is always present when technology is being discussed, and in order for the relationship to continue, not only need new innovations to be presented but also the price for the solutions must be balanced with their capabilities.

In summary, four of the interviewees believe that systematic development of a relationship is possible. The suggested ways for systematic development are generally associated with social exchange, information exchange, mutual trust, and long term mutual commitment. Half of the interviewees consider their relationship with Nordic ID as systematically developed and central element of the development has been social and information exchange that has mostly existed in the form of Nordic ID solving customers' problems, which is found to be effective by the interviewees. The importance of inter-personal contacts is also high in this context. Trust appears to be one of the most important factors affecting their willingness to continue a relationship. General responsiveness, good problem-solving abilities, activity in presenting ideas, inter-personal relationships, and technology were also mentioned. The role of support services in developing relationships will be discussed next.

Even though support services are used in marketing before the purchase, the actual functional role of them becomes evident after the purchase. According to the study, support services are found to be especially important shortly after the purchase when the solution is being installed and set to work.

> *The time after the purchase is critical, because it defines the entire system, whether it is found functioning or not.*

The interviewees were asked which support services offered in their industry they find especially important in general, on the industry-specific level. Several support services were brought up, for example factors associated with installation were found important. Representing a different aspect of support services, technical workshops were brought up by an interviewee. The possibility to participate in the technical workshops where information about new products and also about new partnerships of Nordic ID could be obtained was found extremely useful.

When a relationship already exists, the information-delivery role is emphasized, and that was also the main benefit perceived by the interviewee. Organizing this kind of activity requires special work and attention from the supplier, but can be deemed as an area of rising importance as conducting business in modern, global environments. Besides providing information, elements like workshops can be used as a differentiator from other companies offering similar equipment.

Although several post-purchase support services were mentioned as being especially important and useful, maintenance and repair together with technical support were appreciated most by the interviewees. A possible reason for this is that maintenance and repair together with technical support are often present in critical episodes of the relationship. Single critical episode or a series of them occurring in a relationship affect the development of it, and can have a generative, degenerative, or neutral effect on the development of the relationship. When the effect is positive, additional trust and understanding is created and the relationship develops. Episodes associated with any aspect of support services can become critical and a single critical episode can be definitive to the future of the relationship. However, maintenance and repair together with technical support are typically present in these decisive situations where problems occur and the future of the relationship is defined. These episodes can be decisive as severe cost and inconvenience often follows from failures in handling technical problems.

Whereas above discussion describes the perceived importance of certain support services in general, now the effect of the operating environment on these perceptions is examined. When asked about the effects that business environment or other external factors have on the importance of specific support services, three elements of the environment were mentioned; technological environment, competitive environment, and natural environment. Technological aspect of the operating environment has a major influence in the emphasis of certain services. For example the high inconvenience and cost incurred by down-time sets the emphasis on repair and technical support which are used to minimize down-time in case of a problem. Besides the services that are used when a problem has occurred, also services such as maintenance and user training are appreciated as they contribute to minimizing the frequency and amount of issues.

The importance of constant development and of being able to match and supersede competition highlights the need for the product to allow for further development and upgrades as new technologies become available. Thus development of modifications and upgrades is important for the customers while at the same time offering the supplier an opportunity to combine new ideas with possibilities enabled by new technology. In order to be able to produce relevant innovations that the customers find important and useful, information must be obtained.

In addition to the business environment in which the company functions, also the natural environment in which the device is used affects the support service needs. For example diverse weather conditions in Finland were brought up as a factor demanding special attention when solutions are being designed and produced. In general, profound knowledge about customer's business and associated processes gained by rich and efficient information exchange is needed to be able to effectively match the challenges posed by the factors outside the actual relationship. These factors should be followed regularly, as one specific support service can become the determining one defining the choice between competing suppliers.

Besides the environment, also time is a factor that is unavoidably present when relationships are examined. The interviewees were asked if they think that the importance of some specific support services changes with time. On the basis of the interviews it can be stated, that time and relationship development associated with it is generally found to affect support service needs. The effect of time can be seen as coming either from outside or inside the relationship. The effect coming from outside the relationship is associated with general development of industry and changing requirements of end-customers. When this is the case, the relative importance of specific support services can vary on a short notice and thus a comprehensive and flexible approach to the development of support services is needed.

The effects of time that originate from inside the relationship incur learning and affect various aspects of the relationship and interaction. Some of these changes in the support needs come from the change in the product. As supplier learns about the product, it can be developed over time and the problems can be fixed. Thus, also the emphasis of support services can be moved from keeping the products stable and functioning to moving their functionality forward into direction needed. By gaining experience over time the integrators learn about the equipment they use in their solutions.

As a result of this they need less support in installation, technical support, and also in repair and maintenance. This moves the emphasis in the direction of developing the equipment, thus deepening the relationship with the supplier.

Besides learning inside the individual companies, also improved trust and information exchange that take place when the relationship develops are determinant. They enable the companies to learn about the partner, thus making cooperation more effective and facilitating joint research and development. Cooperation becomes easier to coordinate as both companies know immediately and intuitively which are such points in the development process that are especially important and in which areas their knowledge is of high importance in order for the joint project to succeed.

This study supports the view about the role of support services being bigger after the purchase than before the purchase. For example, installation and technical workshops were mentioned as being important, but maintenance and repair together with technical support were found especially important. This could be because they are often present in problematic situations which may define the future of the whole relationships. Environment appears to affect the companies' support service needs. Complex technological environment of the companies increases the importance of maintenance, repair, technical support, and user training whereas importance of technology together with intense competition was found to highlight the needs for upgrades and modifications. Also the characteristics of natural environment are found to be factors that need special consideration from the supplier.

Besides environment, also time is found to affect support service needs. As the industry develops, new needs arise for the operating companies. Whereas the effect of the development of industry is coming from outside the relationship, also inside the relationship changes in needs take place. Generally, as the companies learn about the products, the emphasis can be moved from repair and technical support to moving the functionality of the products forward by developing modifications and upgrades.

Support services in the relationship with Nordic ID

Whereas previous section has been concentrating on the interviewees' views of support services in their operating industries in general, now their views about the relationship with Nordic ID will be discussed. This is done by defining how the characteristics of their relationship with Nordic ID relate to the industry-level factors that were brought up earlier.

When asked about the role that support services have had in developing their company's relationship with Nordic ID, all the interviewees recognize the role and also find it quite important.

> *I think because it's there, it's probably one of the things that have allowed our relationship with Nordic ID really progress.*

The services found as the most important for relationship development are different from those that are found as the most useful in general. The ones that are considered especially important for relationship development are discussed first. In association with this area of importance of support services, two separate viewpoints can be distinguished. This division follows the grouping of the interviewees to integrators and retailers as it is derived from the purpose for which the equipment is bought. The integrator group generally sees the importance of support services as a factor enabling them to maintain their own customer relationships, whereas the retailer group appreciates the positive effects that they have been able to derive within their organization. This is natural as many retailer companies are customers for integrator companies.

Two specific services were brought up in the group of integrators. Whereas one interviewee describes the whole package of support offered by Nordic ID as well-functioning, some specific services of high importance were brought up by the others. According to an integrator, warranty has had a paramount role in their relationship with Nordic ID. Another specific service that is mentioned by an integrator is customization of products. This was found crucial in order to be able to continue a project with a customer.

In the retailer group two specific services brought up by the interviewees were upgrades and modifications and technical support. The success in the business environment of retailing requires constant technical development in order to hold on to market shares. Technical support, on the other hand, has been used when a specific problem occurs. Technical support is found to be important also in other context than just the development of the relationship, so it will be discussed later in detail.

The support services rated as being the most important ones for their operation include maintenance and repair, technical support, product development and upgrading, and warranty. Generally, it can be said that there are two factors contributing to the importance of maintenance and repair in conjunction with the equipment produced by Nordic ID. Firstly, the equipment is commonly used in large scale logistics and industrial solutions in which a disturbance at some point affects all processes following it. In conditions like this the cost of down-time is substantial, thus emphasizing the need for sufficient maintenance to prevent problems and also for a quick and reliable repair service. Secondly, as the products represent new technology and are relatively complex, they are more prone to problems compared to simpler, tested equipment using older technology. In addition, industry-specific factors further heighten the role of maintenance and repair.

For the retailers the additional importance stems from characteristics of the industry. As the retailing business is fast-paced, in a case of a problem it is important that the failure is repaired swiftly and that the unnecessary problems are disposed beforehand by careful maintenance. As with the integrators, while the solutions supplied by them are used in similarly critical environments, the effect of system-level aspect is added. If one of the equipment used in the solution is causing problems, it hampers the functioning of the whole system. Consequently, maintenance and repair are found to be an absolute necessity by both groups, thus possessing considerable importance in the area of support services. Even though it is difficult to differentiate oneself on the basis of physical repair and maintenance, in many cases they can be definitive factors affecting the decision whether to continue or to terminate the relationship.

Technical support is an important element of support service offering as its functioning can determine how the bought equipment is perceived in the customer's organization. At some point a problem is sure to occur, and if technical support is not able to help it can have a profound effect on the perceived image of the selling company. Thus the technical support needs to be able to produce correct and useful information and answers that genuinely help the customer.

Besides sufficient expertise, another characteristic of well-functioning technical support is that it can produce solutions to customer's problems quickly. Fast provision of solution in a case of a problem lessens the costs and difficulties faced by the customer. The speed of response in technical support is found to be about average or a bit slow compared to the competitors. However, in relation to the resources available, the technical support of Nordic ID is found to be working quite quickly.

The equipments are continuously getting more sophisticated and technologically advanced, possibly increasing future needs for technical support. This together with the increasing amount of units in use world widely is likely to increase the workload of technical support. Thus the resources allocated to it should be increased accordingly in order to avoid negative effects on the functioning of support and consequently on the related company image. Technical support often includes elements provided on-line, as they are cheaper to produce and faster to use than actually having a person to visit the customer who has encountered a problem. On-line elements typically include services such as help desks. Not having a round-the-clock helpdesk was mentioned by one of the integrators as a negative aspect compared to competitors with bigger resources. However, before adding elements such as these, it should be made sure that the resources allocated to technical support are sufficient. Otherwise special characteristics of support will eat up resources of conventional technical support, causing more trouble and costs than benefits.

Personal contacts appear to have a role also in the provision of technical support. The good speed of a response appears to be at least to an extent associated with having direct contacts with the right people inside Nordic ID. As quickness is important in the context of technical support and personal contacts provide quick exchange of information, this reliance on personal contacts is understandable as it can increase considerably the speed by which problems are solved. However, the knowledge and contacts of individual persons inside Nordic ID should be tried to be shared in a way that would benefit the whole organisation. This could even out the workload within the function. Information sharing should be more efficient also in the network of partner companies, and the new solutions should be presented more actively to the end-customers. This would decrease situations in which support is asked for a problem that has already been solved but which was not in the knowledge of other customers. By decreasing such situations, time and effort of both customers and Nordic ID would be saved.

One element of support service offering that is found to be especially useful by two of the retailers is aiding creation of contacts with the third party companies. This is rated as the most important aspect of support services by two of the three retailers. These connections for example with software developers are useful when new functionality is needed. By being directed to a suitable developer, the retailing companies can avoid the cost and inconvenience of having to find a partner able to deliver suitable and compatible solutions. Directing customers to certain developers is also useful for Nordic ID as it makes it possible to maintain ongoing relationship with a customer even though something cannot be offered at the moment. Furthermore, by directing the customers to selected third party companies, some coordination about the customer relationship and the solutions provided can be maintained. Also, as several customers are using similar software solutions, the characteristics associated with the specific solution can be taken into account in product development of Nordic ID, thus enabling it to be more in-control and efficient.

Customizing, user training, documentation, and remote diagnostics were suggested by Nordic ID as potential aspects of support services offering future opportunities and development possibilities to further enhance support service provision. Thus specific questions concerning their importance and current level were asked. Customizing equipment can be defined as producing upgrades and modifications to the product or service to the specification suggested by the customer. This is usually done to increase the products efficiency in use and to make implementation of the product easier. Customization provoked two opposite views thus dividing the interviewees into two. Two of the interviewees found customization of products highly important, even to a degree of it being an elementary prerequisite of using equipment produced by any company in the first place. However, on the other hand all the others stated that concentrating on the development of standard product should be the number-one priority.

When the customization is done on a level that it does not require considerably extra work, as for example with accessories or such, it is found to be justified. But when customization requires extensive product development and technical support work, it is deemed to be wasting of resources of general technical support and product development and consequently unnecessarily increasing their workload and slowing their functioning.

What is wanted, then, is effective technical support and product development work concentrating on the well-functioning core product, design of which it makes it possible for the integrators to customize the equipment to their customers' needs when necessary. This could be achieved by strengthening the division of responsibility so that Nordic ID concentrates on developing the core product, enabling the integrators to concentrate on customizing. This would enable both Nordic ID and the integrators to maximize the effectiveness of work and thus minimize time delays associated with technical problems and still make it possible for the customers to have customized solutions when needed. The organization of co-operation between the integrators and Nordic ID should include a possibility to bring features achieved by customization into the core product, at least to some extent. Customization itself should be encouraged as it is an effective differentiator that can be used in acquiring new customers and binding them to the supplier. In order to do this, however, the cooperation between Nordic ID and integrators needs to be close and controlled, and the integrators' technological abilities become essential. If the equipment is to be designed in such a way that the integrators can customize them, there needs to be considerable amount of trust and social exchange between Nordic ID and the integrators. Again, this needs to be taken into consideration already when partners are selected.

Besides technological customization of equipment, also support services can be customized to suit customer's needs. Again, two different approaches were brought up. According to an interviewee, one should strive to carefully find out what support services are needed most and then keep the support service package standardised in order to avoid problems associated with various and complicated practises. On the other hand, importance of customer orientation was highlighted; when customer has a need, solution to this need should be thought of and then priced accordingly leaving the decision of whether or not to use the service at the given price to the customer. If the company is flexible enough and possesses sufficient resources to provide tailor-made services like this, it has potentially very powerful tool by which customers can be kept close to the supplier, allowing the relationship to be developed. However, the risk of getting mixed in various and parallel service concepts is evident. Also, when the service offering is divided into small elements, the risk of wasting resources by scattering them in several directions becomes present.

Documentation associated with the equipment is essential when complex industrial products are dealt with. The important role of documentation in the business area of Nordic ID is illustrated by following quotes.

> *When there are a lot of end-users, good documentation and good checklist eliminates 60-70 percent of pointless support calls and repairs. So it's very important.*
>
> *Obviously, you've got to have documentation in order to understand how the product works and to help the use of a product.*

Even though the literature suggests that documentation associated with b-to-b marketing is usually quite complete and extensive, on the basis of this study the quality of documentation, timeliness of it, and especially convenient access to it appears to be more important in this context. This could be because the environments in which Nordic ID's products are used generally include various technological solutions used together, so consequently there exist a considerable amount of documentation as all the technology needs to be covered by it. When this is the case, easy and quick access to basic information becomes especially important. Easy and convenient use of equipment resulting in part from useful documentation can contribute strongly to end-users' willingness to continue using the equipment in question, which on the other hand should affect the company's decisions when new equipment and technology is needed. Thus, offering purposeful documentation helps in its part to maintain relationships. Furthermore, when important facts about the equipment are clearly visible from documentation, the decision to purchase the equipment in question is made easier.

World Wide Web is found to be a good channel for delivering documentation, as it addresses the issues of easy and quick availability rather well. The biggest advantages of on-line support are fastness and much lower costs compared to a situation where experts would otherwise be needed to visit the location. The aspects associated with the high speed of information delivery were found interesting.

As the importance of a fast problem-solving is high, remote diagnostics provide interesting possibilities for the future as they can be used to accelerate and intensify the functioning of repair and technical support services. On the basis of this study the benefits derived from remote diagnostics appear to be situation-specific, but at the same time a sound area for the future development nonetheless. According to this study remote diagnostics offer considerably more value to the end-customers than to the integrators, as the integrators in most cases possess enough knowledge about the devices to be able to diagnose them and react accordingly.

Accordingly, remote diagnostics should be designed to be flexible in such a way that also the integrators can use them in supporting their customers.

CONCLUSIONS AND MANAGERIAL IMPLICATIONS

The results of the study indicate that systematic relationship development is possible. However, the systematic aspect appears to be heavily grounded on the factors through which relationships develop over time in the first place: social exchange between individuals, information exchange, mutual trust, and long term mutual commitment. Thus the systematic approach is in using those important factors of relationship development and coordinating them in the desired direction. In order for the relationship to be continued in the first place, several factors are relevant. Trust, open communication, supplier's problem-solving abilities, inter-personal relationships between the individuals of the companies, and also technological capabilities all need to be on a sufficient level.

On the basis of the research, support services can effectively be used in establishing new customer relationships. Their importance for the customer supercedes even the aspects associated with the product itself. When the relationship is already established, the effect of support services changes. Rather than actively developing it, support services keep the relationship active and open for more interaction. It is at this point when information and social exchange together with the trust become especially important, as they ultimately develop the relationship.

According to the results of this study the authors suggest that the relationships and their development together with the impact of support services can be best illustrated as it is shown in Figure 4.

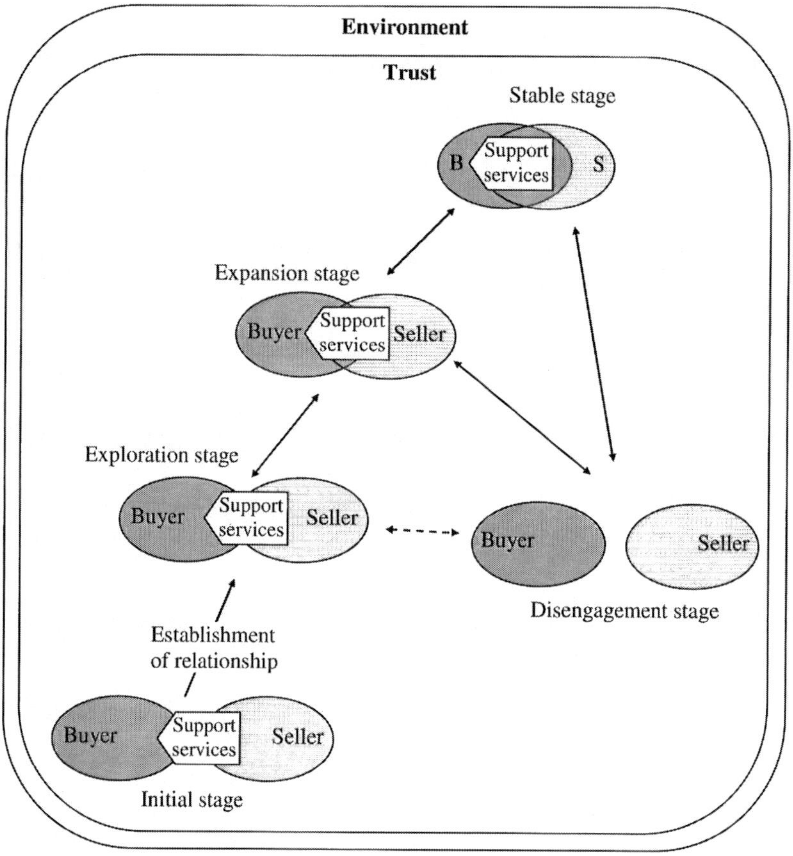

Figure 4. Koistinen & Eskola (2008) model on relationship development and support services

Koistinen & Eskola (2008) model on relationship development and support services illustrates the routes of transition between the development stages *within one relationship and one set of interacting parties*. Therefore, a start of a new initial stage with a new company after consequences management stage with another company is not pictured. For the same reason return to initial stage with one company after negative evaluation of another company in exploration stage is not pictured. The trust is used to describe the emotional setting in which the relationship develops. The arrows marking the transitions between stages describe the increased information exchange that enables the transitions.

According to the results of the study, there can be suggested many managerial implications. The authors argue, firstly, that managers cannot overestimate the meaning of information sharing in support services. Secondly, managers should consider the possibilities to establish an open information sharing within a network of interconnected companies. This network could be built from companies that complement each other's abilities and advantages. Thirdly, development associated with each company's technological core competence could be coordinated by partners possessing necessary knowledge of the subject. This would increase innovativeness and efficiency of research and development. Finally, some responsibility of customization could be moved to the partner possessing most knowledge about the subject, thus lightening the workload of the producer of the core product. Over time the companies in the network could also be able to provide support for the products of the other companies when possible.

REFERENCES

Book

Dwyer, F. Robert – Tanner, John F. Jr. (1999) *Business marketing: Connecting strategy, relationships, and learning.* McGraw-Hill Book Co: Singapore.

Gross, Andrew C. – Banting, Peter M. – Meredith, Lindsay N. – Ford, I. David (1993) *Business marketing.* Houghton Mifflin Company: Boston.

Halinen, Aino (1994) *Exchange relationships in professional services. A study of relationship development in the advertising sector.* Publications of the Turku School of Economics and Business Administration: Turku.

Holmlund, Maria (1996) *A theoretical framework of perceived quality in business relationships.* Swedish School of Economics and Business Administration Research Reports: Helsinki.

Hutt, Michael D. – Speh, Thomas W. (1998) *Business marketing management. A strategic view of industrial and organizational markets.* Sixth edition. The Dryden Press: Fort Worth.

Håkansson, Håkan (ed) (1982) *International marketing and purchasing of industrial goods. An interaction approach.* John Wiley & Sons Ltd: Chichester, England.

Book in Series

Ford, David (1990) The development of buyer-seller relationships in industrial markets. In: *Understanding business markets: Interaction, relationships and networks,* ed. by David Ford, 42-57. Academic Press Limited: London.

Hallén, Lars – Sandström, Madeleine (1991) Relationship atmosphere in international business. In: *New perspectives on international marketing,* ed. by Stanley J. Paliwoda, 108-125. Routledge, Chapman and Hall Inc: New York.

Journals Article

Burger, Philip C. – Cann, Cynthia W. (1995) Post-purchase strategy. A key to successful industrial marketing and customer satisfaction. *Industrial Marketing Management,* Vol. 24, No: 2, 91-98.

Cann, Cynthia W. (1998) Eight steps to building a business-to-business relationship. *Journal of Business & Industrial Marketing,* Vol. 13, No: 4/5, 393-405.

Coulter, Keith S. – Coulter, Robin A. (2002) Determinants of trust in a service provider: the moderating role of length of relationship. *Journal of Services Marketing,* Vol. 16, No: 2, 35-50.

Dwyer, F. Robert – Schurr, Paul H. – Oh, Sejo (1987) Developing buyer-seller relationships. *Journal of Marketing,* Vol. 51, No: 2, 11-27.

Halinen, Aino – Tähtinen, Jaana (2002) A process theory of relationship ending. *International Journal of Service Industry Management,* Vol. 13, No: 2, 163-180.

Heskett, James L. – Jones, Thomas O. – Loveman, Gary W. – Sasser, W. Earl – Schlesinger, Leonard A. (1994) Putting the service-profit chain to work. *Harvard Business Review,* Vol. 72, No: 2, 164-171.

Lam, Shun Yin – Shankar, Venkatesh – Erramilli, M. Krishna – Murthy, Bvsan (2004) Customer value, satisfaction, loyalty, and switching costs: an illustration from a business-to-business service context. *Journal of the Academy of Marketing Science,* Vol. 32, No: 3, 293-311.

Markeset, Tore – Kumar, Uday (2003) Design and development of product support and maintenance concepts for industrial systems. *Journal of Quality in Maintenance Engineering,* Vol. 9, No: 4, 376-392.

Markeset, Tore – Kumar, Uday (2005) Product support strategy: conventional versus functional products. *Journal of Quality in Maintenance Engineering,* Vol. 11, No: 1, 53-67.

Mathieu, Valérie (2001a) Product services: from a service supporting the product to a service supporting the client. *Journal of Business and Industrial Marketing,* Vol. 16, No: 1, 39-58.

Oliva, Rogelio – Kallenberg, Robert (2003) Managing the transition from products to services. *International Journal of Service Industry Management,* Vol. 14, No: 2, 160-172.

Patterson, Paul G. – Spreng, Richard, A. (1997) Modelling the relationship between perceived value, satisfaction and repurchase intentions in a business-to-business, services context: an empirical examination. *International Journal of Service Industry Management,* Vol. 8, No: 5, 414-434.

Ravald, Annika – Grönroos, Christian (1996) The value concept and relationship marketing. *European Journal of Marketing,* Vol. 30, No: 2, 19-30.

Zeithaml, Valarie A. – Berry, Leonard L. – Parasuraman, A. (1996) The behavioural consequences of service quality. *Journal of Marketing,* Vol. 60, No: 2, 31-46.

Proceedings

Schurr, Paul H. (2004) Buyer-seller relationship development episodes: theories and methods. *Proceedings of the 20th IMP-conference,* Copenhagen, Denmark, 2004.

Sutton-Brady, Catherine (2001) Relationship atmosphere. The final chapter. *Proceedings of the 17th IMP-conference,* Oslo, Norway, 2001.

Designing Rapid Services for Competitive Manufacturing

Deepak A. Sachdev, Darius P.K. Singh
School of Engineering
Centre for Technology and Innovation Management
Institute for Industrial Innovation,
Faculty for Design and Creative Technologies
AUT University, New Zealand
Email - darius.singh@aut.ac.nz

Introduction

The purpose of this article is to identify the need for a framework and provide a framework for designing services. The expected outcome is services, such that they are rapid, efficient and effective and therefore a source of competitive advantage for organizations. The performance of an organization is dependent on a combination of its core activities and its linkages with suppliers and customers. It has been suggested (Drucker, 1991) that managerial perspectives and horizons are widening and include communication and collaboration, to exploit opportunities and increase value creation, across functions and companies. Slack, et al (1995) state that "*The design activity in operations has one overriding objective: to provide the type of products, services, and processes which will satisfy the operation's customers.*" As such, exploiting these opportunities encompasses an integrated approach and focus on, sourcing, manufacturing and delivering. Customers; internal and external, expect faster service and would appreciate this assuming other things being equal. In some cases the instantaneous service may even override other considerations e.g. price. Rapid service design is an integrated plan of action to enable rapid service delivery, thereby meeting customer expectations and organizational goals. Through continuous and effective use, across multiple and varied transactions, organisations can develop it into a core competence, thereby enabling sustained competitive advantage.

Research Context and Scope

Manufacturing is essentially a collection of activities to convert inputs into desired outputs based on customer requirements. However, it should be noted that firms within a region/country/location have access to identical resources and are subject to similar environments, thus their competitiveness and performance is based on their individual capabilities and their competitive advantage over other firms. Decision making structure at a typical manufacturing firm is depicted in fig. 1. It highlights the nature of decision making – strategic, tactical and transactional, and the associated roles that make these decisions. The core focus of this study as depicted is at the transactional level of decision making, the study assess the impact of other levels of decision making, i.e. strategic and tactical, on transactional decisions and thereby on transactions themselves.

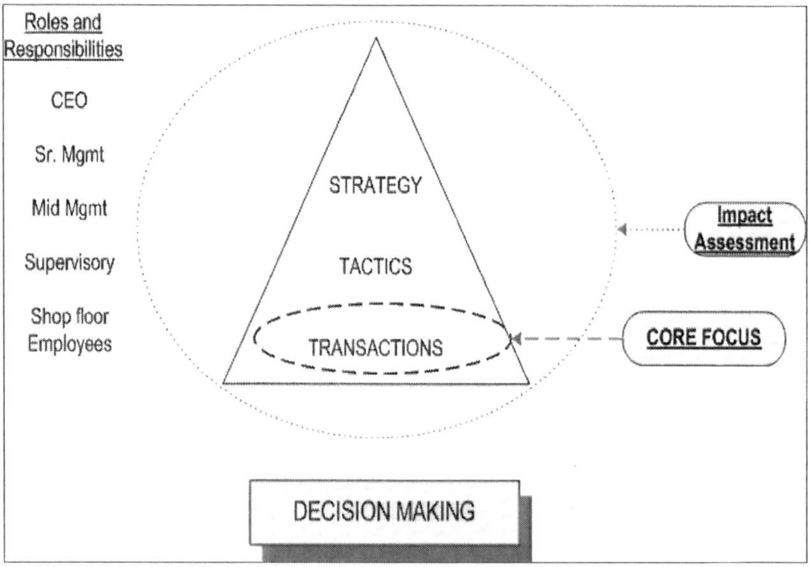

Figure 1. Typical Manufacturing Firm - Decision Making Structure

Transactions are performed through prevalent business processes within an organization. It is during the activities within the process that transactional decisions are made and actions undertaken accordingly. The key cross-functional business process which is employed from point of contact by a customer to the successful quote based on customer requirements can be captured in the Request for Quote (RFQ) process from the customer view point and the Quotation process from the manufacturing organisation's viewpoint. Manufacturing related business practices typically involve many handoffs, multiple information systems, walls within and between organisations, in the RFQ Process. Consider the following scenario "states" as a progression that can be visualised on applying the framework successfully, alternatively this can be seen as the significant contribution of this research, it is worth repeating that the context of the improvement is "speed" observed by the service delivery, in the form of the quote being rapid, which can be measured by improvement in time.

Figure 2. Quote Scenarios – Improvement in Time

Literature Review

The existing knowledge areas that are referred to, and their important insights tailored into an encompassing list of areas to address. This list will then be the reference check of the framework to be developed. As part of this research the knowledge areas to be considered are:

1. Manufacturing systems
2. Technology Management
3. Service Design
4. Competitive Strategy
5. Supply Chain Management

1. Manufacturing systems

Oberoi, et al (2007) study the impact of sourcing and technology on the level of flexibility in manufacturing and its importance of meeting customer demands cost effectively. Their focus of the study relates tactical and strategic levels, through an exploratory study of manufacturing organisations in north India. They highlight that supplier competencies and response times have a high influence on manufacturing flexibility and competitiveness. At its extreme manufacturing systems are being designed to self control work within and between work centers (Schild and Bussmann, 2007). They describe such a self control manufacturing system at DiamlerChrysler and conclude that though it was developed and demonstrated successfully, such a system could not be further rolled out due to investment and financial return constraints. On the other hand, alternative traditional manufacturing systems are attempted to be improved by various practices, for eg. theory of constraints case study at Hitachi Tool Engineering, Japan (Umble, et al, 2006). They mention that this initiative results in an improvement in time based metrics like, inventory turnover, lead time and on-time-delivery. Advanced manufacturing can also enhance competitiveness by a focus on knowledge management (Gunasekaran and Ngai, 2007). Nascimento, et al (2007) describe an approach for product development, through a case study of Embraer, an aircraft manufacturing organisation in Brazil. They focus on the organizational design aspect and mention that a matrix and cell structure facilitates effective decision making. Also highlighted is the need to involve customers and suppliers at multiple stages. Aminpour and Woetzel (2006) discuss an approach for applying lean techniques, in the manufacturing context to factories in China. At a high level the approach, which is equally applicable to any manufacturing organisation worldwide, discussed is identifying and fixing objectives based on customer needs, identifying correct manufacturing processes for meeting these objectives, selecting and training managers who can deploy the required amendments on the shop floor, piloting the changes and rolling out to the entire factory or business unit.

2. Technology Management

It is important for an organisation to have a continuously changing landscape of technology management over the years, with creativity and innovation, to identify new opportunities continuing in radical ways (Larson, 2007). Information needs to be managed as critically as tangible assets. Avenues include; operational, business activities and business intelligence (Beyer, et al, 2008). Information Technology needs to be an integral part of the business process, should also be effective and aligned (Sphilberg, et al, 2007). Technology management needs to have strategic management linkages, should identify core and non-core processes and needs to interpret internal and external information in the context of the firm (Brockhoff, 1996). Vasconcellos, et al, (2007) describe a model that builds on Porter's competitive strategy framework and utilizes it for linkage and planning between technology and strategy, at Wahler-Brazil, an auto parts manufacturer in Brazil with headquarters in Germany. Kaplan and Anderson (2003) demonstrate a simplified and more effective method of activity based costing, the term is time based activity based costing. It is explained by them that this method rests on two key requirements viz. cost of capacity and time taken for an activity. Ravichandran (2006) mentions that reducing a bullwhip effect requires various tactical initiatives; these are highlighted by two case studies of organizations in India, Hindustan Oil Company and Skittles India Limited. Saraph et al. (1989) identify eight key quality factors that are critical, leadership and quality policy, process management, product and service design, supplier quality, quality department, data and reporting, employee relations, training. Farrel et al (2003) highlight the importance prioritizing technology investments and the critical linkage between innovation and diffusion of technology by competitors. Prioritization needs to be such that, technology investments are focused on specific and few areas of importance. Moreover if technology enhancement focus is in areas where competition can easily catch-up due to fast diffusion, then the advantage to the organization is for a limited time, as against application of technology enhancements to those organizational business processes that are different from competition.

3. Service Design

Kotler (2003) states that "all" businesses have an element of service; moreover in today's environment of diminishing areas of competitive advantage, service provides a promising avenue, to gain and maintain ongoing competitive advantage. Ermer and Knipek (1998) demonstrate the use of quality function deployment through its matrices, to achieve service design. Service design thus acquired reflects quality, consistency and meets customer needs. Shostack (1987) demonstrates service design by using a process perspective, through examples involving diagrams, relationships and focussing on the benefit to the customer. Stewart (2003) provides a framework of service design. Central to the method is the focus for improving service quality and utilises elements of task, treatment, and tangibles to design. Richard Rumelt explains in the context of strategic dynamics that "value denials" i.e. products and services that are possible to be supplied and also sought by the customer, but not currently being thought of and thus not being offered by an organization, can be a source of competitive advantage by providing service to the customer uniquely (Lorallo D. P. and Mendonca L. T., 2007, pp 63). Beinhocker (2007 pp5) quotes Albert Einstein "Everything should be made as simple as possible, but not simpler." and states that since markets and companies involve human beings, the interactions are complex and through managing this complexity is the way that organizations create value. Heywood et al (2007) provide a recipe for "cracking the complexity code", which in their opinion constitutes the following areas to address, institutional complexity and individual complexity. Institutional complexity i.e. the number of steps or interactions inside and outside the company for an activity, eg. the steps required to provide service to a new customer. Whereas individual complexity is the level of difficulty in performing individual steps by a person, i.e. how hard they are to do. Beardsley et al (2006) mention that competitive advantage can be gained from interactions by an organization by focusing on being innovative, collaborative and reducing hurdles such as unwanted hierarchies and organizational boundaries. There needs to be a process based view of work and improvement of processes by identifying and removing wasteful activities, thereby improving employee productivity. Also management of interactions on the above lines, especially for tacit interactions i.e. those requiring case based processing eg. could be a customer quote, can create lasting competitive advantage. The result is that employees are more effective in their work as against being efficient in wasteful areas or processes where potential efficiency improvements are marginal.

4. Competitive Strategy

Drucker (2001) states that the purpose of a business is to create a customer, profit motive merely validates business decisions in this journey and the key function deriving from this purpose is ensuring that, customer needs are met. Porter (1985) describes his concept of a value chain, value being the worth of an organisations offering to the buyer and chain involving the primary and secondary activities of the organisation. Primary activities include inbound logistics, operations, outbound logistics, sales and marketing and service. Support activities include procurement, human resource management, infrastructure and systems. He highlights that competitive advantage can be gained by either cost or value advantage over competitors. Helms and Ettkin (2000) discuss time based competition and elaborate on time as a strategic focus for competitive advantage. Sukumar (2006) also highlights the importance of time, by emphasising the need for fast thinking and action, for sustainable competitive advantage. Kotler (2003) highlights that a great company creates competitive advantage by cultivating a unique mix of capabilities, over its competitors. He further states that strategy can be viewed as successfully providing a different set of benefits to targeted customers and thereby meeting their needs. Strategy, execution, culture and structure are mentioned as key differentiators for superior results (Nohria, et al, 2003). Ghemawat (2002) provides a historical explanation of competition and business strategy and takes us though a journey through time, focussing more on the developments from 1950 to recent times. Roxburgh (2003, pp35-36) encapsulates the essence of strategic thinking. Strategy is the activity of identifying unique sources that give a competitive edge; therefore a strategist needs to be on the look out for novel and unusual ways in which this can be achieved. This may require looking at the periphery to seek out areas for innovation. Also important is to avoid the herding instinct.

5. Supply Chain Management

Supply Chain management can be defined as (Simchi-Levi D., et al, 2004):

"... a set of approaches used to efficiently integrate suppliers, manufacturers, warehouses and stores so that merchandise is produced and distributed at the right quantities, to the right locations, and at the right time in order to minimize system wide costs while satisfying service-level requirements." The issues to be addressed are in the following areas, Network planning (flow of goods for distribution), inventory control, supply contracts (pricing and volumes), distribution, integration and partnering, outsourcing, make or buy decisions, product design, customer value (need fulfilment and contribution to customers), information technology, decision support systems and reverse logistics (Simchi-Levi D., et al, 2004). Walters (2002) noted that no one function can operate in isolation and that intra-inter organisational cooperation is essential. A study by D'Avanzo (2003) examined 623 global companies across 24 industries in order to assess performance towards supply chain excellence. The author placed these companies in four categories, based on analysis of financial information over the periods of 1995-1997 and 1998-2000: Leader (superior performance), Transformer (improved performance), Decliner (deterioration), and Laggard (non-superior performance). The study identified the four key factors that distinguished leaders from the laggards. In particular it indicated that the company viewed their supply chains as critical drivers of shareholder value and competitive differentiation. The companies also endeavoured to incorporate supply chains in their business strategies by building innovation in their operating models and rigorously executing their supply chain strategies. By so doing the companies were able to adapt their practices to meet changing customer requirements. An APQC (2005) best practice report on tradeoffs for inventory and service levels highlights the importance of credible demand information, coupled with a passion for applying and executing strategy, across organisational levels, cross functionally.

While the above individual areas have a treasure of research already done and ongoing, there is no discernable evidence of research, specifically on deployable frameworks of service design for competitive manufacturing, which integrates all the above theoretical areas. Thus there is a need for rapid service delivery at transactional activity level, and ideas from above especially, Kotler (2003), Simchi-Levi D., et al (2004) and Oberoi, et al (2007), can be developed further and used to develop a deployable framework of rapid service design for competitive manufacturing.

Framework Development

Researchers have identified that the future of research in operations will involve addressing real problems and developing the theory to support managerial decision making in those contexts (Chopra et al, 2004). The goal of a business is to have sustained profitability. It can achieve this through meeting customer needs and meeting them more effectively than its competitors. However having a competitive advantage on a long term basis requires differentiation in business offerings, compared to other players in the field. Differentiation could be in terms of price, quality and time.

Considering a manufacturing business, price and quality are dependent on the core resources of the manufacturing activities, e.g. equipment, raw materials, scheduling, employees skills etc. These can be bridged by various competitive players over short and medium term, though access and suitable duplication of identical resources. However, other things being equal, the element of "time" demonstrated by rapid response, is relatively more difficult to duplicate as it involves, an effort towards a major portion of a business areas, i.e. Sales and Marketing, Production Planning, Production, Procurement, etc. Thus developing a core competence around time has the potential to provide sustained competitive advantage. The problem is "how" can this be achieved? For example consider a scenario for developing a framework in the context of rapid prototyping, the typical context and focus would be the manufacturing process itself, as depicted in the following figure:

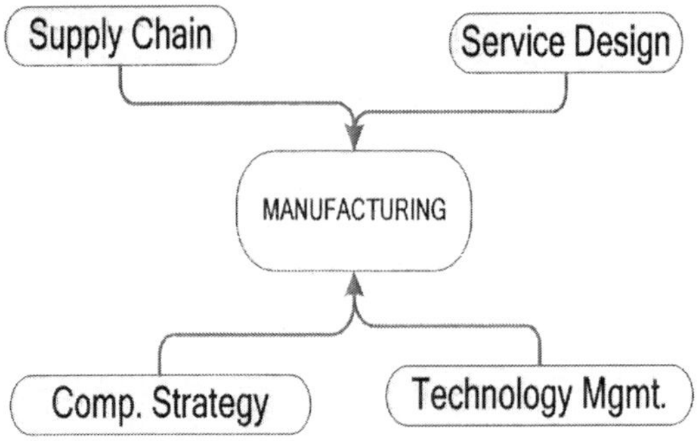

Figure 3. Inward Perspective: Manufacturing Focus

The Service perspective is conversely an Inside – Out view, i.e. the focus is on Service design, from the customer point of view. The framework developed essentially looks at the transactions within a “process” involving and revolving around the customer needs and integrating the essentials of Supply Chain, Technology Management, Manufacturing, with Service Design as its core as depicted in the following figure:

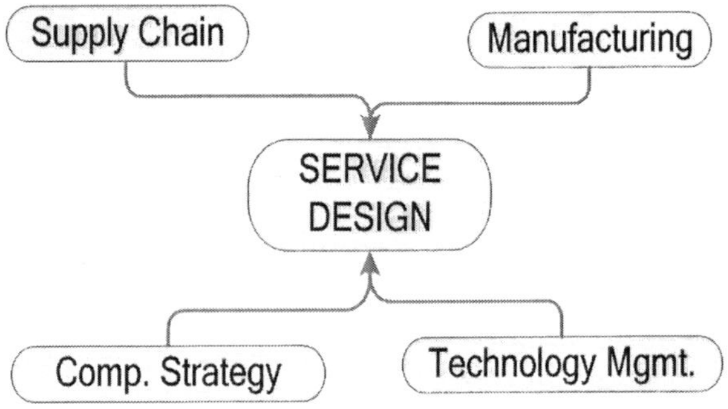

Figure 4. Outward View: Service Design Focus

Thus a framework for service design should thus ideally encompass innovativeness and collaboration including elements of; alignment of manufacturing operations, technology management by its use to deliver optimal processes, competitive strategy by meeting customer needs better than competition, supply chain management by satisfying service-level requirements. Further consider a hypothetical case of a trading organisation, the successive scenarios with the Service design framework applied; successively reflect the increasing effectiveness in the sequential scenarios. The case considered here is for a quotation process, requiring a made-to-order manufacturing process (Supplier).

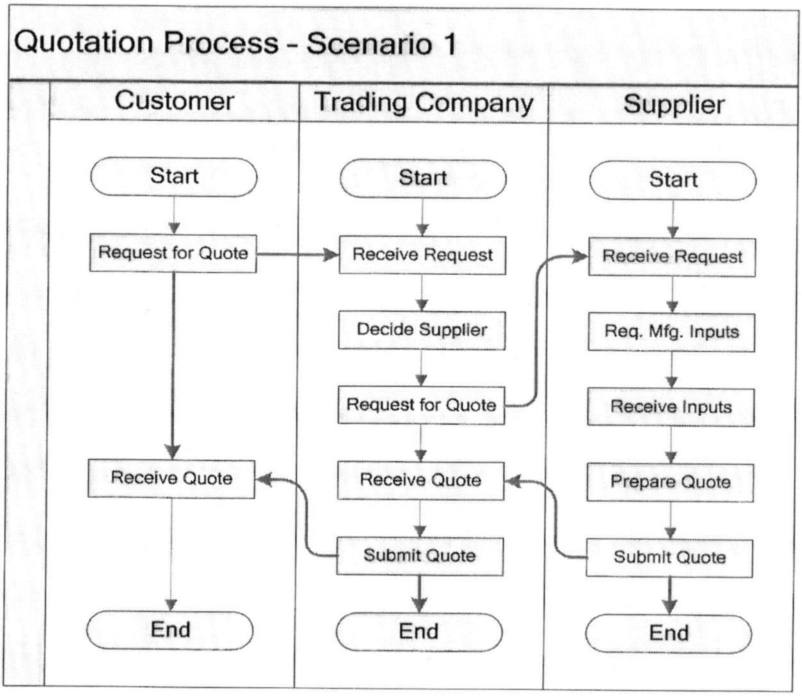

Figure 5. Quotation Process - Scenario 1

Interfaces for Scenario 1

- Customer requests trading company to quote
- Trading company requests marketing department at supplier i.e. manufacturing organisation. Note that trading company may itself have marketing and sourcing departments, eg. overseas office, say in United Kingdom and sourcing office could be in India, this may mean identical activities at Trading company, similar to the Supplier.

- ➢ Supplier marketing department requests its manufacturing department for inputs in quotation.
- ➢ Supplier manufacturing department provides inputs to marketing department.
- ➢ Marketing provides quotation to trading company
- ➢ Trading company provides quotation to customer

<u>Interfaces for Scenario 2</u>

- ➢ Customer requests trading company to quote
- ➢ Trading company requests marketing department at supplier end, this step is not required in this scenario.
- ➢ Trading company requests the manufacturing department directly for quotation
- ➢ Supplier marketing department requests manufacturing department. for inputs, this step is not required in this scenario.
- ➢ Supplier manufacturing department provides inputs to marketing department; this step is not required in this scenario.

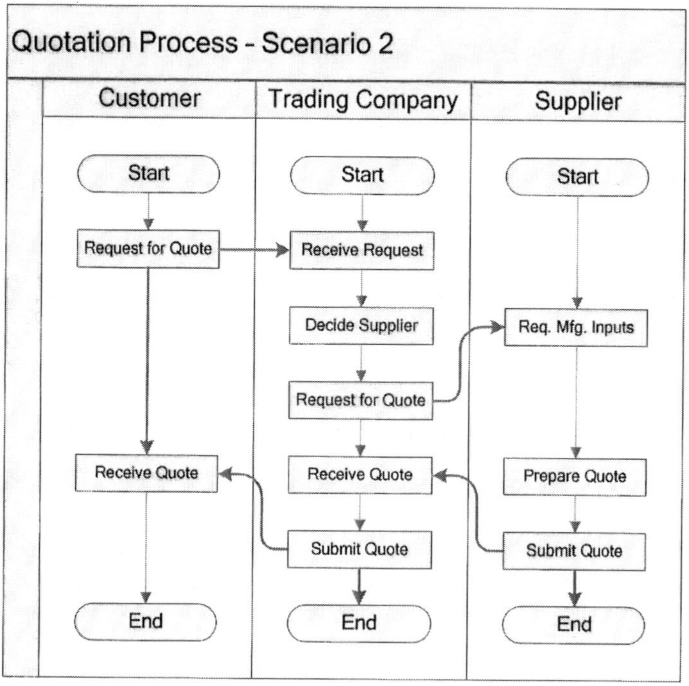

Figure 6. Quotation Process - Scenario 2

- Manufacturing department directly provides quote to trading company
- Trading company provides quote to customer

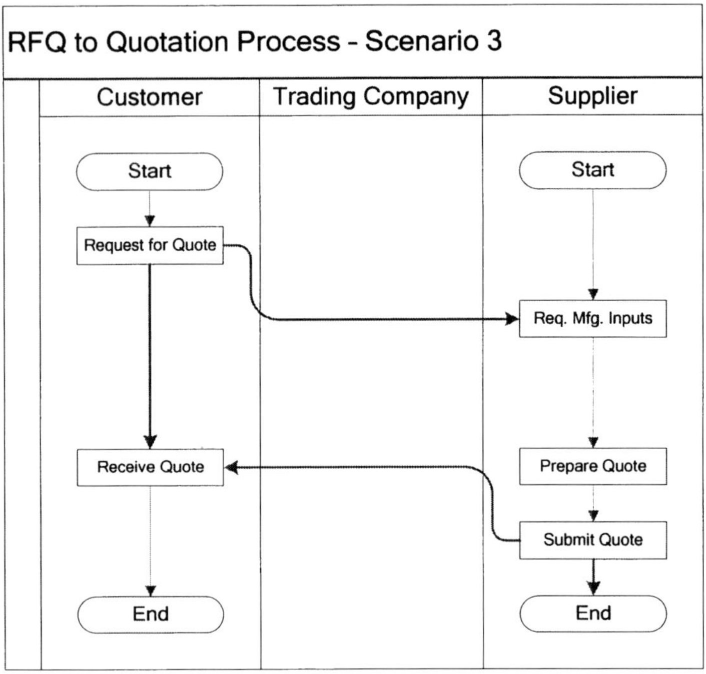

Figure 7. Quotation Process - Scenario 3

<u>Interfaces for Scenario 3</u>

- Customer requests the manufacturing department directly for a quote.
- Manufacturing department directly provides quote to trading company.
- Customer requests trading company to quote, this step is not required in this scenario.
- Trading company requests marketing department. At supplier end, this step is not required in this scenario.
- Trading company requests the manufacturing department directly for quotation; this step is not required in this scenario.
- Supplier marketing department requests manufacturing department for inputs, this step is not required in this scenario.
- Supplier manufacturing department provides inputs to marketing; this step is not required in this scenario.

- Manufacturing department directly provides quote to trading company, this step is not required in this scenario.
- Trading company provides quote to customer, this step is not required in this scenario.

Discussion and Conclusion

The scenario 1 thus describes a host of more activities, as compared to scenario 2; where as scenario 3 has even fewer activities. The ideal state would be scenario 3 to achieve relative rapidness in the quotation process. It should be noted for scenario 2 that the flow is directly from the customer to marketing at the trading organisation and further to the manufacturing department at the supplier end. Thus the marketing department at the supplier end is by-passed. Whereas in scenario 3 the flow is directly from the customer to the manufacturing department at the supplier end, thus by-passing both the trading organisation and also the marketing department at the supplier end.

The manufacturing process remains unchanged, for the quotation under consideration. The potentially disproportionate, actual time spent in core manufacturing and the other allied activities which constitute the overall service for the customer, also should be considered is. Note that this desirable shift from scenario 1 to scenario 3, requires achieving the characteristics to have a smooth transition, this requires supply chain collaboration, suitable contracts and developing of trust, for trading company to be bypassed and yet get due value of the business or ensuing transactions, and allowing the trading company to focus on more value added activities, e.g. increasing business development or engaging new customers, etc.. Also it should be emphasised that while the current discussion is for quotation process only, it can be extended to the other service processes like invoicing, etc. yielding dramatic lead time and associates cost reduction, as can be visualised through comparison of the three scenarios above, this in turn will make the organisation more competitive, through two fold effects of being more effective and desirable to deal with, thus gaining an competitive advantage.

The framework developed for Service Design enables rapid services, thereby providing an avenue of sustainable competitive advantage. It draws on the areas discussed in the literature review; it has an inside-out view focussing on the customer. It is worth repeating, (Drucker, 1991) that managerial perspectives and horizons are widening and include communication and collaboration, to exploit opportunities and increase value creation, across functions and companies.

REFERENCES

Aminpour S. and Woetzel J. R. (2006) "Applying Lean Manufacturing in China" *Mckinsey Quarterly*, Special Edition – Serving the new Chinese consumer, pp 106-115.

APQC (2005) "Inventory Optimization: Balancing the Asset versus Service Tradeoff", A best practice report from APQC Publications.

Beardsley S. C., Jhonson B. C. and Manyika J. M. (2006) "Competitive Advantage from Better Interactions" *Mckinsey Quarterly*, No. 2, pp 52-63.

Beinhocker E. D. (2007) "Managing in a Complex World" Mckinsey Quarterly No. 2, pp 4-5.

Beyer Mark A., Friedman Ted, Young Colleen M. (2008) "Fundamentals of Information Management for the CIO" *Gartner Research,* ID Number: G00155475, 2008.

Brockhoff Klaus (1996) "Technology Management in the Company of the Future", *Technology Analysis and Strategic Management,* Vol. 8, No. 2, 1996.

Chopra, S., Lovejoy W. and Yano C. (2004) "Five Decades of Operations Management and the Prospects Ahead", *Management Science*, Vol. 50, No. 1, January 2004, pp. 8-14.

D'Avanzo R. (2003) "The Reward of Supply-Chain Excellence", Issue 22, *Optimize Magazine.* December, Available: http://www.optimizemag.com/article

Drucker P. F. (1991) "The Coming of the New Organization", *Change Management - Harvard Business Review*, paperback No. 90064, 1991.

Drucker P. F. (2001) "The Essential Drucker" Harper Collins Publications, Inc.

Ermer D. S. and KNIPEK M. K. (1998) "Delighting the customer: Quality function deployment for quality service design" *Total Quality Management*, Vol. 9, Nos. 4&5, S86-S91.

Farrel D., Terwilliger T., Webb A. P. (2003) "Getting IT Spending Right This Time" *Mckinsey Quarterly*, No. 2, pp 118-129.

Ghemawat P. (2002) "Competition and Business Strategy in Historical Perspective" *Business History Review* 76 (Spring 2002): 37-74.

Gunasekaran A. and Ngai E. W. T. (2007) "Knowledg management in 21st century manufacturing", International Journal of Production Research, Vol. 45, No. 11, 1 June 2007, 2391-2418.

Helms M. H., Ettkin L. P. "Time-based competitiveness: A strategic perspective" *Competitiveness Review;* 2000; 10, 2.

Heywood S., Spungin J., Turnbull D. (2007) "Cracking the Complexity Code" *Mckinsey Quartely*, No. 2, pp 84-95.

Kaplan R. S. and Anderson S. R (2003) "Time Driven Activity Based Costing" *SSRN,* id 485443, Nov.

Kotler P. (2003) "Marketing Insights from A to Z", John Wiley & Sons, Inc.

Larson Charles. F. (2007) "50 Years of Change in Industrial Research and Technology Management", *Research*

Technology Management, January – February 2007. pp 26-31.

Lorallo D. P. and Mendonca L. T. (2007) "Strategy's Strategis: An Interview with Richard Rumelt" *Mckinsey Quarterly* No. 4, pp 56-57.

Nascimento P. T., Vasconcellos E. and Lucas P. C (2007) "Organization for Product Development: The Successful Case of Embraer", *Management of Technology – Vol. 1*, Challenges in the Management of New Technologies, edited by Horlesberger M., El-Nawawi M., Khalil T., pp 311-322.

Nohria N., Joyce W., Roberson B. (2003) "What Really Works", *Harvard Business Review*, July 2003' pp 42-52.

Oberoi J.S., Khamba J. S., Sushil and Kiran R. (2007) "The relative impact of technology and Sourcing practices in managing manufacturing flexibilities – Evidence from large and medium scale enterprises in India" *Human Systems Management* 26 (2007) 199-215.

Roxburgh C. (2003) "Hidden Flaws in Strategy", Mckinsey Quarterly, No. 2, pp 26-39.

Saraph, J., Benson, P. and Schroder, R. (1989) "An instrument for measuring the critical factors of Quality Management", *Decision Sciences Journal*, 20, pp. 810-829.

Schild K. *and* Bussmann S. (2007) "Self-organization in Manufacturing Operations", *Communications of the ACM,* December 2007/Vol. 50, No. 12.

Shostack G. L. (1987) "Service Positioning through Structural Change", *Journal of Marketing* Vol. 51 (January 1987), 34-43.

Simchi-Levi D., Kaminsky P. and Simchi-Levi E. (2004) *"Managing the Supply Chain – The Definitive Guide for the Business Professional"*, The McGraw-Hill Companies: New York.

Shpilberg David, Berez Steve, Puryear Rudy and Shah Sachin (2007) "Avoiding the Alignment Trap in Information Technology", *MIT Sloan Management Review,* Fall 2007, Reprint Number 49102.

Stewart D. (2003) "Piecing Together Service Quality: A Framework for Robust Service", Production and Operations Management, Vol. 12, No. 2, Summer 2003, pp 246-264

Sukumar S. (2006) "Creating a Strategic Measurement System", *Infosys White Paper,* Infosys.

Umble M, Umble E. and S. Murakami S. "Implementing theory of constraints in a traditional Japanese manufacturing environment: the case of Hitachi Tool Engineering", *International Journal of Production Research*, Vol. 44, No. 10, 15 May 2006, 1863-1880.

Vasconcellos E., Monterossi J. O., Bruno M. A. C. (2007) "Identification of Technological Threats and Opportunities: Conceptual Framework and Application to an Autoparts Company" *Management of Technology for the Service Economy: The 16th International Conference on Management of Technology*, May 2007, Miami, Florida, pp 1346-1362.

Walters D., Halliday M. and Glaser S. (2002) "Added Value, enterprise value and competitive advantage", *Management Science,* 40/9, pp 823-833.

PART 4

" Industry "

- *Productivity assessment of implementing wireless technologies in steel construction using simulation technology.*
- *Next generation PLM - an integrated approach to development of hybrid products in the telecommunication industry.*

PRODUCTIVITY ASSESSMENT OF IMPLEMENTING WIRELESS TECHNOLOGIES IN STEEL CONSTRUCTION USING SIMULATION TECHNOLOGY

AMINE GHANEM
Assistant Professor of Civil Engineering and Applied Mechanics, California State University Northridge, Northridge, California 91330-8347; PH (818) 677-3891; FAX (818) 677-5810
email: amine.ghanem@csun.edu

Abstract

Steel construction is considered as a process that involves many related activities. Pre-engineered buildings (PEB) steel parts are required to be installed in a specific order due to structural safety requirements and to the logical sequence of erection. However, shipping, transportation, unloading and on-site storage do not take into account the erection order of the assembly. As a result, considerable time is consumed locating, sorting, and identifying steel components. Integrating promising information technologies such as radio frequency identification (RFID), mobile computing devices and wireless technology can be useful in improving the effectiveness and convenience of information flow in construction projects. Current information and communication technology may be incorporated in the operational process for efficient assembly at the job site. For models implementation, two steel construction projects were used as examples to represent an overview of the current practices of existing steel operations, find out potential productivity problems and sources of waste, and explore the potential possibilities for improving current processes. An information flow diagram followed by a simulation model for Pre-Engineered processes from shipping the steel materials until erection on the construction site was developed. Then, the improved steel process is modeled and the productivity of both models is compared by using simulation software (Micro Cyclone). Feasibility study of implementing the proposed project tracking model and its impact on the construction project is conducted. Implementing the proposed model will result in at least 7% of additional profit for the first project, and 5.5% in case of the second project. Those results were only based on direct savings.

KEYWORDS: IT, Micro Cyclone, Pre-Engineered Building, Simulation, Productivity, Wireless technologies

Introduction

Construction is one of the largest industries in the United States, and it is the second largest employer after government agencies (U.S. Census Bureau 2004). However the construction industry has suffered from low performance due to low productivity, high accident rates, late completion, and poor quality (Kashiwagi et al., 2004). A study performed by Nuntasunti (2003) summarized five factors preventing the construction industry from improving performance: 1) Fragmentation of the construction industry; 2) project specific nature of construction; 3) temporary nature of relationships; 4) competitive bidding system; and 5) stand alone islands of communication. In particular, the steel construction industry faces significant challenges to remain competitive (Ricles 2002). As a result, decreasing fabrication and erection time for steel frame buildings, while increasing the safety of workmen during construction are issues that must be addressed. According to the American Institute of Steel Construction (AISC), a 25% reduction in time required to erect a steel frame structure is needed. One commonly cited means of overcoming labor shortages and improving productivity, cost effectiveness, and competitiveness is through the use of advanced technologies such as information technology (De la Garza et al., 1998). Pre-engineered buildings (PEB) steel parts are required to be installed in a specific order due to structural safety requirements and to the logical sequence of erection. However, shipping, transportation, unloading and on-site storage does not take into account the erection order of the assembly. As a result, considerable time is consumed locating, sorting, and identifying steel components. For models implementation, two steel construction projects were used as examples to represent an overview of the current practices of existing steel operations, find out potential productivity problems and sources of waste, and explore the potential possibilities for improving current processes.

Steel Construction Projects

The first construction project is Turbocor project, a 65,000 sq. ft. home for a Canadian-based air-conditioning compressor manufacturer located in Innovation Park, Tallahassee, Florida. Construction began November 2005 and was completed in July 2006. Sperry & Associate Company, based in Tallahassee, FL, was the design build firm responsible for the project. The total project cost was approximately $3 million and was under the budgeted amount. The project included the construction of one huge building divided into two parts: the first part represented the A/C plant, and the second part represented the offices. The building was (PEB) structure with metal sheeting as wall and roof covers.

The second construction project is Jefferson County High School, located in Monticello, Florida off U.S. Highway 19. Construction of the high school began in June 2003 and was completed in July 2004. The Haskell Company, based out of Jacksonville, FL, was the design-build firm responsible for the project. The total project cost was approximately $12.6 million and was under the budgeted amount. The project included the construction of nine total buildings. These buildings were also (PEB) structures with Metal-Stud- Crete panel walls.

Model the Existing Steel Construction Processes

Based on field observations and information collected on the construction sites, and on interviews with key players, a diagram of materials and information flow was formulated to represent the flows of information and materials throughout the fabrication, shipping, and erection phases. This diagram provides guidelines of what information the steel crew needed to perform a specific task, how data were shared, and where to get those data from. In addition to that, it helped us to diagnose existing PEB steel construction processes to find out alternative processes. Figure 1 presents the materials and information flow of the PEB steel construction processes. As seen in the diagram, information and materials are passed from one party to another via conventional approach. Key data sources within the defined system are: fabricator, worker, foreman, and site manager. The next two sections will define the productivity of the current practices and need for improvement.

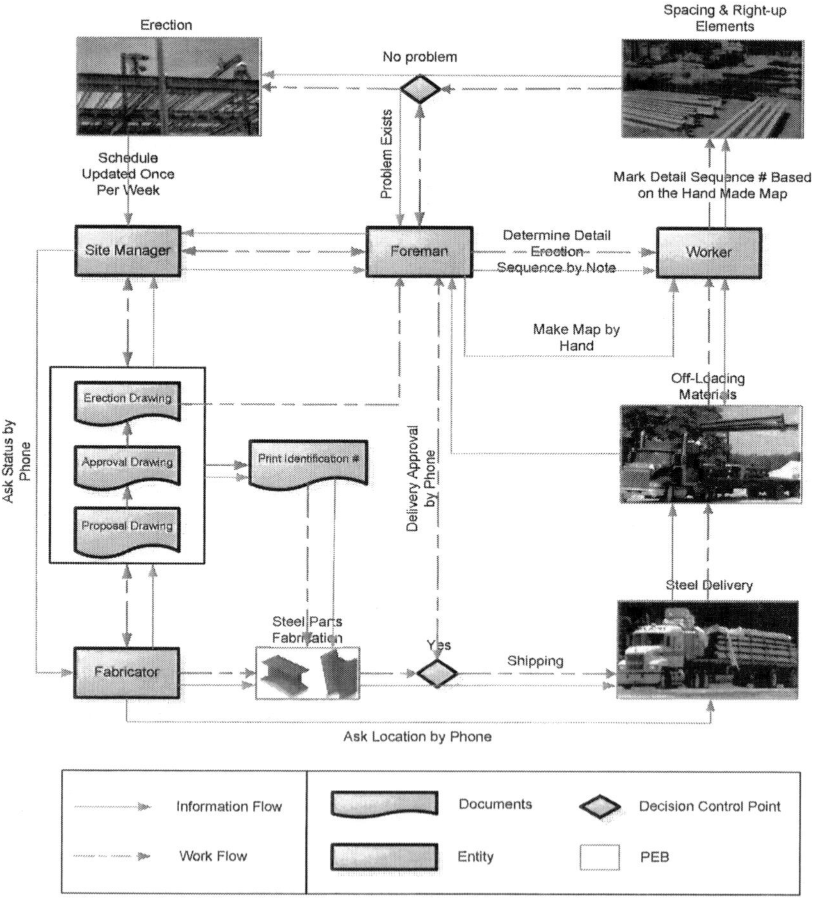

Figure 1. Materials and Information Flow

Productivity Measurement

Spending time on the construction site, observing tasks and movements of the steel crew allowed the authors to collect productivity ratings data for PEB steel operation. It was not possible to observe and analyze all of the tasks. Therefore, sample measurement was taken for different processes and it was then incorporated into a simulation model to measure work productivity.

Oglesby proposed three classifications for productivity ratings. The classifications are: effective, contributory, and ineffective (Oglesby et al., 1989). The percentages of crew time in all categories of the shakeout, unloading and erection operations are summarized in table 1. The percentage rate was calculated by dividing the observed number in each category by the total observation.

It is noted that only 32% of total crew time was spent on the effective work for the shakeout, unloading and erection operations while the contributory work took 33% of crew time. The ineffective work accounts for 35% of crew time on the average. Average time spent walking empty-handed, idle, non-work related communication, searching for materials, and break time are respectively 5%, 10%, 9%, 5%, and 4%. These kinds of activities must be the areas to be addressed for productivity improvement.

On the other hand, continuous time study involved the measurement of time required to address the specific job under present conditions. Among all process, we were interested into 4 of them: shipping, unloading, shakeout, and erection operation. Those constitute one cycle in our study. It was proposed that continuous time study be made of an average of three cycles for unloading, shakeout and erection. The next section presents the implementation of our study into a simulation model to study the cycle productivity.

TABLE 1 PRODUCTIVITY RATINGS

Categories	**Items**	**Percent of Total Number (Shakeout, Unloading & Erection)**					
		Foreman	Erector1	Erector2	Worker1	Worker2	**Average**
Effective Work		10%	36%	29%	30%	33%	32%
Contributory work	Communication	33%	14%	10%	15%	15%	17%
	Searching for materials and tools	2%	20%	15%	9%	9%	11%
	Read document	20%	0%	0%	0%	0%	4%
	Subtotal	55%	34%	25%	24%	24%	33%
Ineffective work	Walking empty-handed	6%	52%	7%	3%	3%	5%
	Idle	4%	7%	10%	15%	17%	10%
	Correcting error	0%	0%	0%	0%	0%	0%
	Non work related communication	20%	4%	5%	9%	3%	9%
	Searching for materials	2%	8%	6%	6%	8%	5%
	Break time	2%	6%	1%	3%	3%	4%
	Non identifiable	1%	0%	25%	10%	9%	9%
	Subtotal	35%	30%	17%	46%	43%	35%
	Total	100%	100%	100%	100%	100%	100%

PEB Simulation Model

Micro Cyclone is used in this study in order to study the construction steel processes. The elements of Micro Cyclone, originally developed by Halpin, were used to model and simulate PEB steel operations (Halpin and Riggs, 1992). The Cyclone system has been used frequently to model construction processes. This frequent use is due to the ability to provide a quantitative way of viewing, planning, analyzing, and controlling the processes and operations. The model formulated took into consideration the delivery and the availability of the steel parts on the construction site. In addition to that, in case there was a problem with erection or with the steel parts, the back and forth communication between the construction site and the steel fabricator was taken into consideration. Based on a lot of trial and error and trying different simulation models, the authors decided to separate both the shipping simulation process from the erection simulation process. The main reason behind this decision was the time shipping process takes with respect to the erection time, and when placing both processes into one simulation model the results looked inaccurate. Figure 2 and 3 presents an example of the PEB steel simulation model for both shipping and erection. The shipping model is formed of 10 entities covering one cycle of the process. The cycle starts from loading the materials at the fabricator shop till offloading material on construction site. The erection model is formed of 30 entities covering one cycle of the process. The cycle starts from steel parts available on the constructions site till erection of one bay of the steel structure. Once the graphical model was established the next step was to transform it into an input language that Micro Cyclone understands. The unit of time used for the input model was hours. Some entities that represented inefficiency in the processes were included in the model. Web Cyclone is case sensitive and all typing should be done in upper case letters. In the simulation output, some resources were idle when others were in use. Next section discusses those entities which represent opportunity for improvement, increase in efficiency and decrease in waste for different processes. The simulation output will be discussed, once the proposed flow diagram is established and its simulation model is performed.

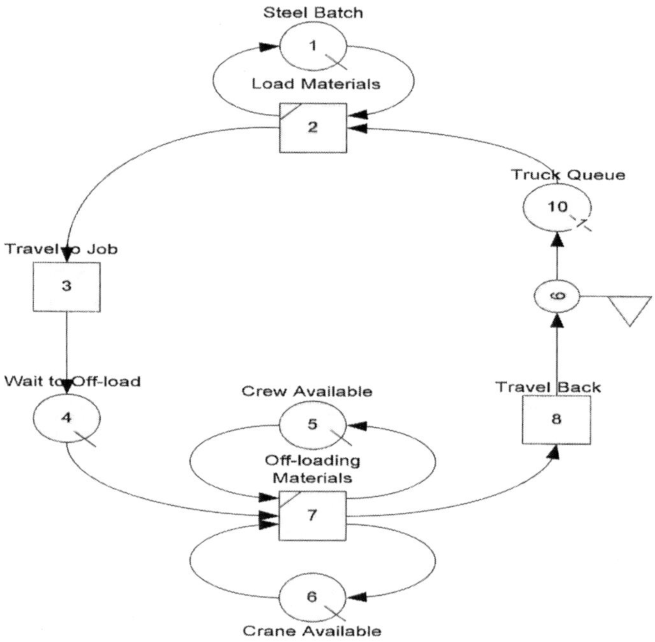

Figure 2. Shipping Simulation Model

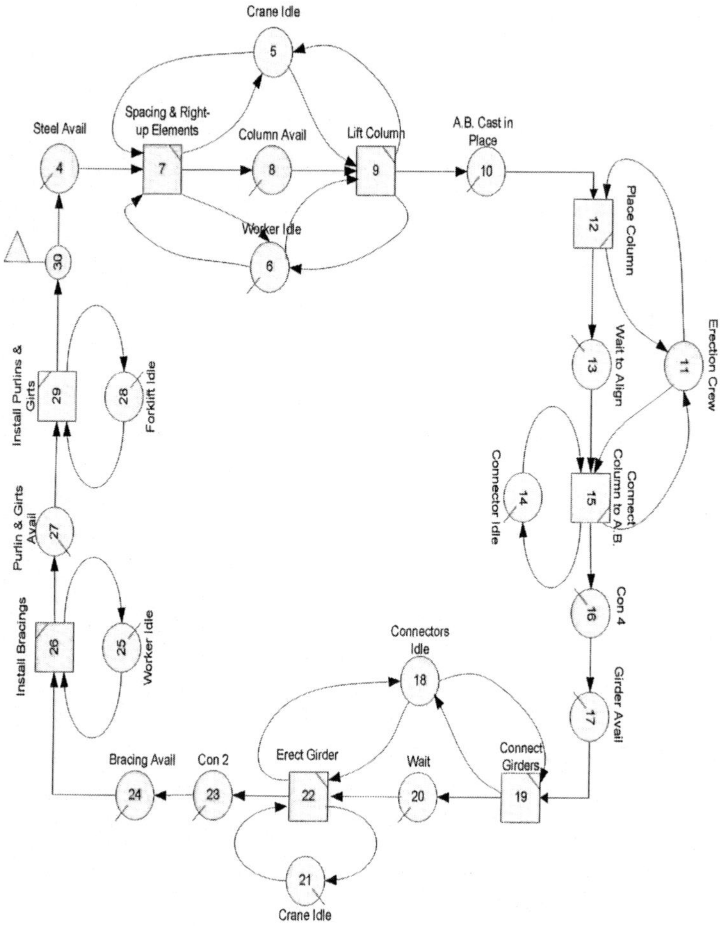

Figure 3. Erection Simulation Model

Process Inefficiency

Once fabricated, manufacturer labeled each steel member with a unique piece mark and sequence number to identify it directly on the erection drawings. However, to make it easy to find materials for erection, workers marked each piece one more time with white chalk based on the erection hand map they made. This process leads to considerable unproductive duplication.

Instead of setting the steel directly off the delivery trucks, all the steel was off loaded and shakeout was done as the steel was

delivered. This practice resulted in double handling of materials in the erection operations.
Another concern was raised when workers tried to locate the exact steel piece to be erected. They had to identify the steel components using paper-based information. As a result, a significant portion of time was spent in the lay-down areas searching by hand to identify components.
During our interview with the people on both construction sites, it was found that material and information flow can be lost, disconnected and distorted while flowing from information sources to end user. Good examples were given to the authors in these two case studies. Workers didn't know when next shipping date was scheduled. Connectors had no full idea of where each steel element was positioned. Foreman had no idea of the status of a shop drawing approval after implementing some changes and which sequence of steel elements was fabricated and stored at the factory and ready to be shipped.
Also it was noticed that critical information from the field to the office had a delay, and lacked real time touch. According to the productivity ratings data mentioned in table 1, foreman spent most of his time communicating with workers, 33% of his time on work related communication and 20% for non-work related communication. These percentages are higher than the average values, 17 % and 9% respectively.
Another two communication problems identified with the current processes were with the approval drawings and request for information. Actually usual delays associated with the steel supply process were encountered during approval stage of shop drawings. Many weeks were wasted due to movement of hard copy drawings from one party to another.
In case a problem related to steel materials aroused on the jobsite and can't be resolved at the field office, site engineer had to submit a request for information (RFI) to the technical engineering department. RFIs were sent via fax with a sketch and a reference in the drawings and in case photos were attached they had to be mailed by federal express. This would delay RFI turnaround times.

Steel Construction Process Updated

A significant number of problems were identified in the previous section resulting from inaccurate data transfers as well from delays and interruption in information flow, thus leading to a wasteful operation and inefficiencies in some processes. So a new approach needed to be developed to not only ensure control of information in a timely manner, but also increase level of communication between multiple processes units for structural steel construction. The main purpose of this section is to present the improved processes on the construction site after implementing the proposed model, and at the same time to present the benefit of implementation by comparing the outputs of both simulation models.

Real Time Project Tracking Model

Construction materials may account for 50-60% of the total cost of a construction project of the elements that comprise the constructed facility. In industrial and heavy construction, mostly prefabricated objects, such as structural steel, pre engineered building and precast concrete elements, are used and installed on site. If the right material could be easily and precisely tracked, found, and distributed to the right location at the right time, tremendous benefits would ensue. Reducing unsuccessful searches for materials would reduce wasted supervisory time, crew idle time, and disruptions to short interval planning, and hence help to increase labor productivity, reduce materials stock piles, and reduce materials management manpower (Bell & Stuckhart 1986). As stated before the authors were interested in three main processes: fabrication, shipping and erection on the construction site. Each steel part has and RFID tag attached to it to give it a special ID that will be connected to a certain activity in the Master Schedule. The tag includes holders with two slots that accept a tie-wrap plastic fastener to attach the tags to the steel parts. Tags can be attached to the steel parts using plastic ties and double sided tapes. There may be as many as 2,000 fabricated steel parts in an average size of buildings.

The truck loaded with steel materials arriving at the project job site, is sent to one of the designated laydown areas. Workers are provided with handheld computer/ RFID reader. The reader used has a form like a standard PC card and can be incorporated into the existing handheld computer using a PC card expansion pack. Omni antenna and software are also installed in the handheld computer. So workers at the laydown area identify the pieces using RFID readers, and coordinate the unloading and storage of the steel parts. Any discrepancies between the packing list and the steel parts unloaded from the truck is sent back to the fabricator for resolution, or in case there are discrepancies between the bill of materials and the unloaded materials, it will be directly captured and fabricator will be contacted. At the same time information related to the installation procedure of the materials under study can be acquired from a central database system. All these information are sent instantly as soon as it is entered to the database server, through the wireless connectivity available on the construction site. All activities have cost account codes to easily relate data to these activities.

Individual objects scheduled for arrival on the construction site are tagged at the vendor factory using radio frequency identification tags. The encoded information is scanned directly into a portable computer and wirelessly relayed to a remote project database. This database has dual functions. A database query returns graphical representations (e.g. computer aided design CAD) like where to install the materials and the method used for installation. Savings are captured when the simulation model for the improved process is performed, especially with the improvement in productivity. Ghanem et al. have an illustrated example of the system implementation.

Development of a data flow diagram

In order to identify the benefits resulting from implementing the proposed model for structural steel construction processes, a proposed data flows diagram was determined based on the field observation. In the data flow diagram as shown in figure 4, key issues that were addressed include real time piece tracking, retrieval of information related to the steel parts, like its location on erection drawings, position and its orientation, and updating the construction schedule.

When the steel component's identity can be read directly through a portable RFID reader, the erection sequence and position information of each piece in their final location are automatically extracted from the common database through wireless networks in real time. This situation makes information stored in the database accessed at any time by people on the construction site, and updates make construction activities from the jobsite available to all process units who need the information in real time. Another simulation model was developed for the proposed model, taking into consideration the new sequence and duration of each activity in the process (Ghanem 2007).

Simulation Outputs

The main output of simulation model is the productivity result, which shows the number of cycles per unit duration. This productivity data is what can be used to evaluate the performance of a process design at a glance. A site engineer can also use the sensitivity analysis function to try to improve the productivity of a process by changing the resources. Based on Web CYCLONE, the construction simulation result of a repetitive cycle in both projects was presented as the productivity. For computing productivity of both shipping and erection process, 'Hours' was used as the time unit as shown in table 2.

The cycle number in Table 2 represents the number of time both systems cycle. For instance, one cycle of shipping steel materials (Refer to Figure 2) includes Queue node (1: Steel batch) to Queue (10). In case of erection process (Refer to Figure 3), Queue node (4: Steel Available) to accumulator (30) is included in the one cycle of the system. It is usually defined by user. Therefore, the

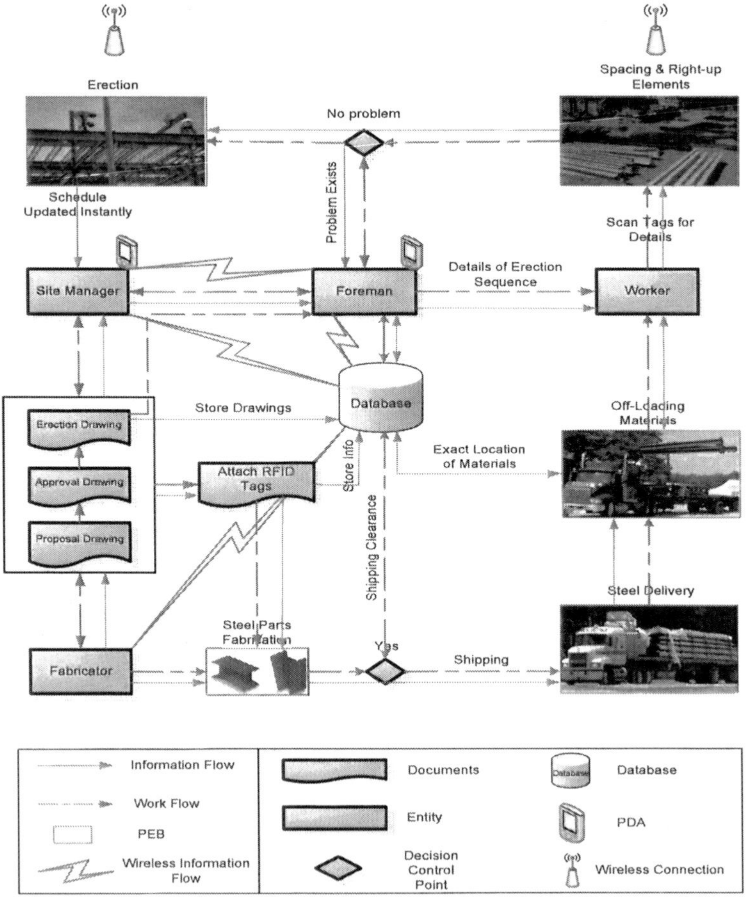

Figure 4. Proposed Materials and Information Flow

productivity of shipping can be calculated through dividing 'Cycle number' by 'Total simulation time' (i.e., 30/107.3 = 0.279) and 300 / 2155.9 = 0.139 for erection). The same is done. It takes into consideration the new sequence and duration of each activity in the process.

TABLE 2 SIMULATED PRODUCTIVITY RESULTS

	Total Simulation Time (unit)	Cycle No.	Productivity/ Time (unit)	Production/ Cycle	Productivity
Shipping	107.3	30	0.279	1 truck	0.279 truck/hour
Proposed Shipping	67.1	30	0.447	1 truck	0.447 truck/hour
Erection	2155.9	300	0.139	1 bay	0.139 bay/hour
Proposed Erection	1699.3	300	0.176	1 bay	0.176 bay/hour

Based on the simulated productivity table, it takes 3.58 hours in order for one truck to reach the construction site using the conventional process. On the other hand it will take 2.24 hours in order for one truck to reach the construction site. At the same time it takes 7.19 hours to erect one pre engineering steel bay. But using the proposed process it will take 5.68 hours to erect one bay. As summary, there will be saving in money at least one hour in the shipping process and 1.5 hours in the erection process. Next section will quantify those benefits and turn it into dollars.

Model Benefits

The objective of this section is to analyze the values of the integrated wireless technologies model functions. The first step was to establish the cost of the technology that was used in this research and this was achieved by conducting a life cycle cost analysis (LCC).

Life Cycle Cost Analysis

The study period for this LCC was taken for one year, because benefits were calculated for one year period. Discount rates and inflation rates were estimated as average figures. Some numbers were based on the previous research done by Singletary (2006). A summary of the LCC analysis is shown in table 3 to cover both case studies. All costs have been discounted back to the beginning of the first month using adjusted monthly discount rate. Planning and feasibility costs are assumed to be incurred at the same time as equipment purchases. The total present worth of life cycle cost for each alternative is shown. As we can notice, the cost of the model for case study is more expensive than case study 1, and that can be explained by the fact that case study jobsite is much bigger, its location is different from case study 1. Next step is to conduct a cost benefit analysis and break even analysis.

TABLE 3 LCC SUMMARY

Name	Value	Units	
Nominal Discount Rate	12.00%	%	
Constant Inflation Rate	2.25%	%	
Adjusted Monthly Discount Rate	0.81%	%	
System Life Cycle	12	Months	
	Case Study 1	Case Study 2	
	Tablet-Wireless Subscription	**Tablet-Wireless Subscription**	**Month Incurred**
Initial Costs			
Planning & Feasibility	$1,800	$1,800	0
Hardware	$17,280	$35,499	0
Software	$10,791	$10,791	0
Wireless Communications	$105	$245	0
Training	$3,326	$3,326	0
Wireless Consultant	$152	$152	0
Total Initial Costs	**$33,454**	**$51,813**	
Recurring Costs			
Hardware Maintenance	$125	$292	1-12

Software Maintenance	$1,049	$3,909	12
Wireless Consultant - Network Maintenance	-	-	3,6,9,12
Wireless Subscription	$240	$560	1-12
On-Going Training	$448	$448	6
Downtime Due to Device Failure	$727	$1,205	1-12
PW Recurring Costs	**$15,049**	**$29,249**	
Total PW LCC	**$48,046**	**$81,062**	

Benefits Quantification

Benefits were divided into four different functions: Site Inspection, Problem Solving, Sensing & Data Access, and E-Document Management. Cost benefit analysis, sensitivity analysis, and/or break even analysis were performed when appropriate for each function.

The first function addresses the need of three common types of inspection: 1) structural, 2) material, and 3) safety. The second function represents problem related to inaccurate drawings and structural pieces result in costly problems because they commonly require significant time and effort to find solution (e.g. RFI process). With this function, field personnel will have direct access to any information from offsite experts who have more experience, or able to solve design problems or other errors, using technology integrated communication tools. The third function deals with the PEB processes inefficiencies identified in the previous chapters, and how superintendent, workers, and erectors wasted a lot of their time to find the right steel parts to be erected, or getting more details of a steel connection drawings, or even being up to date with steel delivery to the jobsite and which section in erection will be next. Not able to find the materials to be used costs more than inconvenience. Time spent searching for assets eats into productivity, and hence profitability. The last function addresses bottleneck in the shop drawing approval process of structural steel

construction. The submittal log maintained five shop drawing packages between Sperry and Associates and American Building Company during the shop drawing approval stage. Using e-drawing will accelerate the process of drawing moving from one party to another and the work flow will be much faster. So if a modification is required, it can be completed on line and sent to the next party involved in the approval process in real time manner.
The benefits occurred from each function independently are added all together to get the total money the model will bring to the construction site. The summary of all benefits is shown in table 4.

TABLE 4 SUMMARY OF BENEFITS

Functions	Case Study 1	Case Study 2
Site Inspection	$7,603	$9,801
Problem Solving	$59,904	$74,956
Sensing & Data Access	$187,094	$667,094
E-Document Management	$7,250	$7,250
Total Savings	$261,851 - $48,503	$759,101 - $81,062
	$213,348 (7.11%)	$678,039 (5.5%)

The total savings for case study 1 is $213,348 which represents roughly 7% of the total cost of the project. For case study 2 is $678,039 which represents roughly 5.5% of the total cost of the project.

Discussion and Conclusions

A significant number of problems in the steel construction industry were identified in this research resulting from inaccurate data transfers as well from delays and interruption in information flow, thus leading to a wasteful operation and inefficiencies in some processes. Wireless technologies can be used to improve the accuracy and timeliness of the data collected from sites. They have the potential to solve this communication problem, increase collaboration, and provide new capabilities through evolving technologies. Two case studies were used to model existing steel construction. A feasibility study of implementing the proposed model and its impact on the construction project was performed. Different statistical analyses were performed to quantify the total benefits. Companies interested in implementing this model within their organization must recognize the nature of the new technologies involved, their life cycle, and most importantly how to integrate their work processes with these technologies. The proposed model should be applied in more real life case studies to generalize the conclusions.

References

Bell, L.C., and Stukhart, G. (1986) “Attributes of Materials Management Systems”. Journal of Construction Engineering and Management, ASCE, 112(1), 14-21

De la Garza, J.M., Howitt, I., (1998) “Wireless Communication and Computing at the Construction Jobsite”. Journal of Automation in Construction, 7(4), 327-347

Ghanem, A. (2007). Real time Construction Project Progress Tracking: A Hybrid Model for Wireless Technologies Selection, Assessment, and Implementation. Dissertation submitted in fulfillment of Ph.D. at Florida State University

Ghanem, A., and AbdelRazig, Y. (2006) “A Framework for Real-Time Construction Project Progress Tracking”. Proceeding of the 10th International Conference on Engineering, Construction and Operations in Challenging Environments, Houston, Texas

Halpin, D. W. and Riggs, L. S. (1992). “Planning and analysis of Construction Operations”. New York: John Wiley & Sons

Kashiwagi D., Sullivan K., Greenwood D., Kovell J., and Egbu C. (2004). “Source of construction industry instability and performance problems”. Performance based studies research group, Arizona State University

Nuntasunti, S. (2003). “The effect of visual based information logistics in construction”. Ph.D. dissertation, department of civil engineering, North Carolina State University, NC 27695

Oglesby, C. H., Parker, H. W., and Howell, G. A. (1989). “Productivity Improvement in Construction”. New York: McGraw-Hill

Ricles, J. (2002). "Next Generation Steel Structures". Presentation delivered at the NIST/AISC
Automated Steel Construction Workshop, Gaithersburg, MD

Singletary, M. (2006). "Assessing the Financial Feasibility of Implementing Wireless Technologies for Construction Management". Master dissertation, College of Engineering, Florida State University

U.S. Census Bureau, 2004 Economic Census, www.census.gov

NEXT GENERATION PLM - AN INTEGRATED APPROACH FOR THE DEVELOPMENT AND MANAGEMENT OF THE PRODUCT SERVICE SYSTEMS IN THE TELECOMMUNICATIONS INDUSTRY

JULIUS GOLOVATCHEV
Head of PLM Competence Team
Detecon International GmbH / Deutsche Telekom Group
Oberkasseler Str. 2
D-53227 Bonn
julius.golovatchev@detecon.com

OLIVER BUDDE
Research Institute for Rationalization and Operations Management
at Aachen University
Pontdriesch 14/16
D-52062 Aachen
oliver.budde@fir.rwth-aachen.de

Consistent definition, categorization and operationalization of product service systems delivered to the markets and customized for different segments is one of the major challenges for the Product Lifecycle Management (PLM) in the telecommunications industry. Each lifecycle encompasses the processes, information, support systems, culture and people involved in delivering the related business functions. It can be argued that process supporting solutions like WFMS, PDM etc. exist today for the creation of a seamless environment for accessing, manipulating and reasoning about product information that is being produced in fragmented and distributed environment. A holistic approach that is suitable for aligning the Product Lifecycle Management (PLM) business requirements to the potentials of new technology driven concepts in the information management like SOA as well as to the environmental and economic aspects over the complete product lifecycle is nevertheless still missing. The authors give reasons why the holistic view on the product lifecycle management is critical for an efficient PLM and present an integrated approach for developing of product service systems compatible with the requirements of sustainability and profitability for the business. In this paper four building blocks of Next Generation PLM are introduced: 1) Product Meta Model; 2) PLM-Process; 3) PLM IT-Architecture; 4) Lifecycle Value Management concept for steering the PLM process on a strategic level like, among other things, by considering portfolio effects. The paper closes with the presentation of a short case study validating the proposed concept for a company in the telecommunication industry.

Keywords: PLM, Product Lifecycle Management, Product Service System, Integrative Approach, Lifecycle Value Management, Product Meta Model

Motivation

As Shark (2004) postulates, Product Lifecycle Management (PLM) is the activity of managing a company's products across the complete lifecycle, from the early stages of conception to the final disposal or recycling of a product. From this definition, a strong interrelation between the value creation process and the PLM of the company can be deduced. Taking the current globally changing business environment into account the PLM can be considered as a strategic weapon for enabling the company to provide an additional value to customers and thereby gain a competitive advantage over their competitors. Especially in the telecommunication industry, which experienced a significant change in its environment during the last five years, PLM is considered to be essential for facing ever shorter innovation cycles (Wöhler-Moorhoff et al., 2004).

In the following two sections, we introduce the main business drivers, which currently have the most impact on a company's business model and are expected to do so in future. For each business driver specific PLM-relevant challenges can be derived. These challenges will be pointed out in detail in the last section of this chapter and will serve as motivation for our integrated PLM-approach in chapter 2. The validation of this approach is the goal of chapter 3 by introducing a case study.

Business Drivers

Business Driver One: Management of the Extended Product

The dichotomy between tangible goods and immaterial services is subject to change according to Saaksvuori (2004). In the service industry as well as in the traditional industries like manufacturing this development towards a product service system can be observed. (Figure 1) Following the argumentation of Mateika (2005) a critical success factor in the manufacturing industry is the ability to provide more profitable services in addition to the tangible good. But this combination of tangible and intangible attributes in terms of the product service system increases the complexity in managing the Product Lifecycle (Saaksvuori, 2004).

Especially in the telecommunication sector products typically consist of several modules that in sum create the customer benefit (e.g. Triple-Play consisting of voice, data and entertainment services). Each module embodies its own lifecycle, which implies an additional product complexity in terms of module design, module management and module removal from the product.

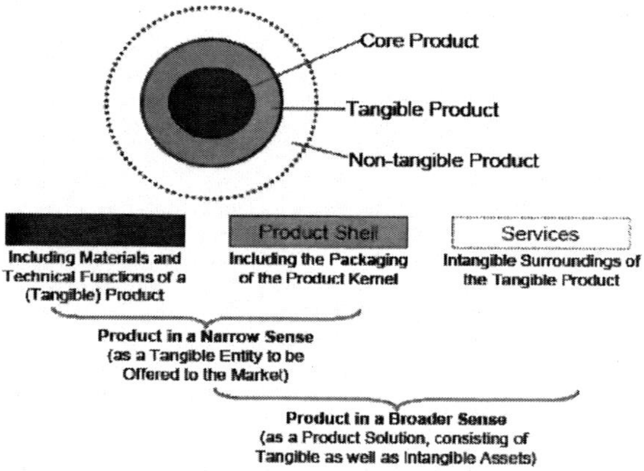

Figure 1. Definition of product service system according to Thoben et al., 2001

Business Driver Two: Globalization

Global and deregulated markets result in global competition that every company is confronted with. For the service industries this results into the shift from the suppliers' market to a buyers' market, where the customer is the focal point. This business driver forces companies to follow either a cost-leadership strategy or a differentiation strategy on a global scale according to Porter (1985). In order to perform in the cost-leadership role, companies have to implement efficient processes regarding the development of complex products and to handle the order management process. Simultaneously companies have to cooperate with numerous business partners, which results in a high effort for coordination. A product differentiation strategy implies processes that ensure short time-to-market on one hand and an effective degree of freedom in the process definition for staying innovative on the other hand.

Business Driver Three: Legal Frameworks

Increasing regulations such as safety, environmental and product reliability will influence company's PLM- process. Especially the resulting effort for ensuring traceability will have an impact. Traceability has two dimensions in the PLM-context. First of all it is related to the PLM process itself. Effective measures have to be in place in order to ensure the flow of transactions to identify where material misstatements due to error or fraud could occur (Sarbanes Oxley Act). Further, traceability in the order-delivery process is about tracking of an individual product units or even components. Legal frameworks like TREAD Act for the automotive industry in the USA or REACH for the chemical industry in EU can be subsumed under this dimension. Companies that need to respond to this kind of regulation must ensure that the implementation and the integration in the existing processes happen to be simple and at the least possible lifecycle costs.

Challenges

Based on the outlined business drivers, challenges for companies can be derived on the strategy, process and IT layer. An overview of these challenges is given in the following table.

Table 1. PLM-relevant challenges (Golovatchev, J. and Budde, O. (2007)

	Strategy Impact	Impact on Processes	Impact on IT
Business Driver 1	Less vertical integration, more horizontal integration of the value chain Focusing on product modularization and utilisation of product platforms to enable the re-use of components	Collaborative PLM-process with defined interfaces Process readiness for the management of complex products	Data exchange standards IT-representation of complex products Appropriate fulfilment of the information need
Business Driver 2	Commitment to either a product individualisation or product standardisation strategy	Improvement in the process automation Inter-company wide process standards	Enterprise Application Integration with external business partner Support of the process execution by providing appropriate IT-functionality
Business Driver 3	Comply to regulations e.g. Sarbanes Oxley Act, TREAD Act	Support for traceability of products by standardised and well documented processes Support for traceability of processes	Management of product related information during the whole lifecycle

Integrated PLM- Approach

The integrated PLM- Approach consists of four components as depicted in figure 2.

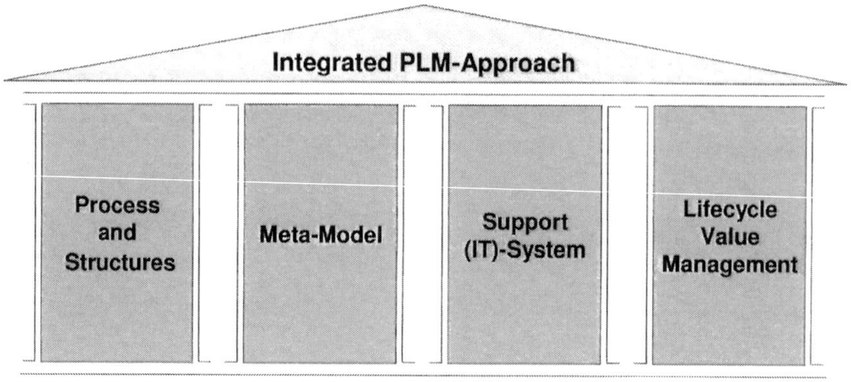

Figure 2. Four components constituting the integrated PLM-approach

In the following paragraphs the different components are described further.

1.1 PLM Meta-Model

Many present quality deficiencies originate from a diffuse definition of products as well as from the inconsistent view on the object "product". The hybrid product should be clearly and precise defined to be developed, commercialized and understand from the customer. There should be a constant product definition with the overall product lifecycle and for all business process.

Definition, Categorization of Products

The product as the main subject of the process needs to be defined and this definition of product should be taken from the customers´ perspective. A product is defined as anything that can be offered to a market for attention, acquisition, use of consumption that might satisfy a want or a need. It includes physical products, services and information (compare figure 1). A product is an entity the customer perceives in its entirety and is characterized by being offered to a market place. The precise definitions of the complete product range and their categorization still remain a problem especially in the service industry (e.g. Telecommunication, energy, financial services industry etc.). For example, in the telecommunication industry essentially four categories of products can be differentiated: market product (service), standard solution, customized (individual solution) and bundle. All offered products and services can be assigned to one of the mentioned categories.

Product definition is not simply a reflection of the engineering design. It also includes the entire set of information that defines how the product is designed, manufactured, operated and managed on the market and finally withdrawed from the market when it becomes obsolete. This product definition is continually updated throughout the entire lifecycle from idea generation until withdrawal from the market and it is the same for all business processes. This view allows considering the product as a core information object (CIO).

Standardization of Products and Modular Product Database

Definition of the product as a CIO is the basis for standardization and modularization of products in the service industry. Product standardization changes the process landscape and reduces costs. Modular product data structure ensures linking the sales perspective to the internal (production) perspective in relation to the offered products and services. The main aim of introducing a modular product structure is the optimization of the product development.

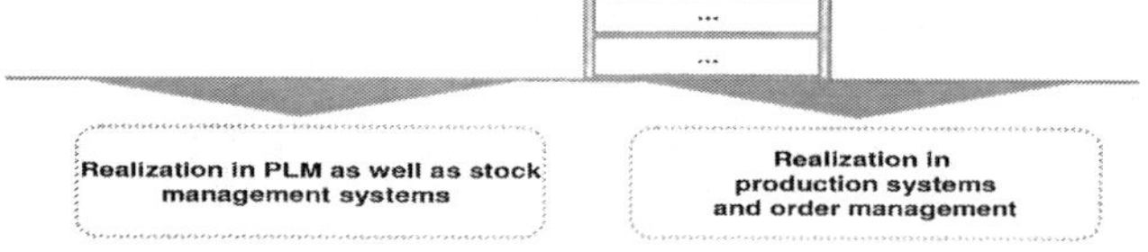

Figure 3. Example for modular product data structure model

The product database and the corresponding processes should be implemented by considering the existing processes, company structures, and cultures. Requirements from the affected functional areas flow into the definition of the product data structure. One master product data management fits all needs of the main processes and its interfaces.

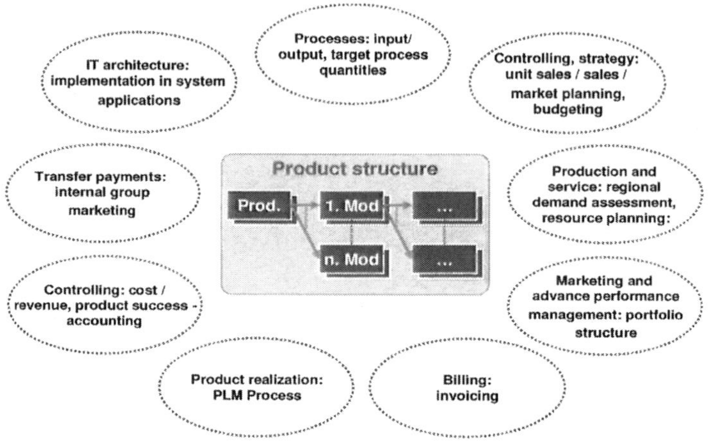

Figure 4. Definition and requirements on product data structure

Using modular product structures leads to simplification and cost optimisation of "product engineering" in the service industry.

1.2 PLM IT-Architecture

The foundation for moving a product through its complete lifecycle beginning with the product idea and ending with the product removal from the market is a solid PLM IT-Architecture, that is customized for the company-specific PLM- requirements. In our perspective, such a PLM IT-Architecture must support the PLM- process in the dimensions: (1) Decision support, (2) Operational support and (3) integration of supplemental business applications. A standardized off-the-shelf PLM-System is therefore not the tool of choice as Ausura and Deck (2007) point out. Instead we suggest to rely on a PLM IT-Architecture that re-uses, respectively customizes existing IT-components as far as possible. In the next paragraphs a description of the essential IT-components is given that build up the PLM IT-Architecture.

Decision Support System (DSS)

The main purpose of a DSS is to gather and consolidate data from operative systems in order to provide the senior management with aggregated information regarding the product lifecycle. The Computer Aided Selling (CAS) module provides functionality for the product configuration[12] and product pricing. The component Strategic Resource Management focuses on the long-term resource capacity planning on a strategic level. Product portfolio management aspects are covered with the last Value Management component.

Process Support System (PSS)

On the operational level the PLM- process execution is supported by the PSS. The Workflow Management Systems (WFMS) enables a higher degree of process automation. Especially in the context of distributing and releasing unstructured content like a product specification in cross-functional teams, the WFMS plays an essential role through a strong link to the Product Data Management System (PDM). The Product Modeling System and the PDM are closely coupled. The former defines the product structure, in which types of modules the product is decomposed into[13]. In the database context this functionality is similar to the schema definition. The PDM-systems stores all product relevant data according to this definition and provides different views for each stakeholder e.g. marketing and engineering. The Multi-Project Management System as well as the Collaboration Tools are instruments for managing the product in different phases in a collaborative environment.

[12] The CAS module is also essential in the operative context for the order management. In our understanding CAS goes beyond the operative support and helps to steer the variant management in order to cope with the increasing product complexity nowadays.

[13] For example hierarchies of assembly groups can be defined. Additionally constraints and rules can be modeled constricting the combination of specific modules.

Integration of supplemental business applications

A PLM IT-Architecture is supposed to hold the promise of seamlessly integrating and presenting all information produced throughout all phases of a product's life cycle to everyone in an organization, along with external business partners. For ensuring this functionality, an EAI-approach has to be implemented. According to recent research activities a Service Oriented Architecture (SOA) is most suitable to integrate the business applications from external partner e.g. suppliers as well as integrating the own business applications like ERP, CRM etc. in order to fulfil the promise of seamless integration for becoming a real-time enterprise (Abramovici, 2002).

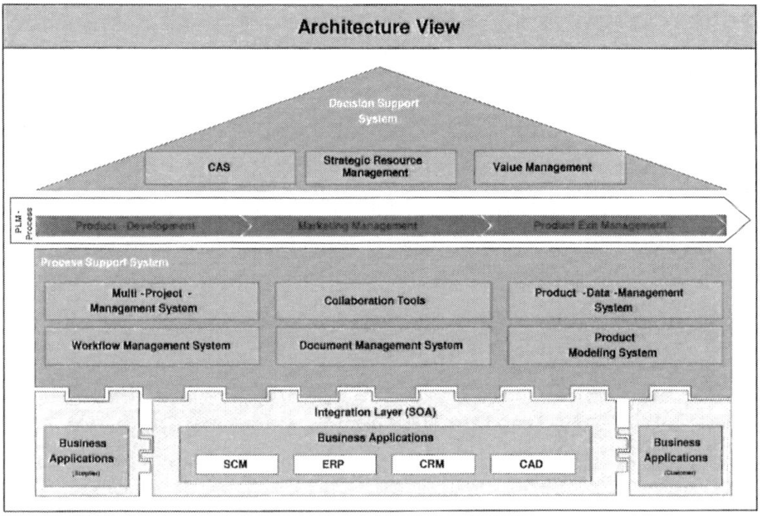

Figure 5. PLM IT- Architecture building the IT-foundation of the PLM-approach (Golovatchev et al, 2008)

1.3 PLM-Process

In this section the PLM-Process description will be introduced. The main design goals of this approach is to facilitate the execution of this collaborative process (efficiency goal) and equally important to align the activities with the strategic goals of the company (effectiveness goal).

Our PLM-Process approach is based on three principles: Stage Gate Approach, Multi-Perspective, Metrics based Management. These principles are subject for further explanation in the following paragraphs.

Principle One: Stage-Gate Approach

The Stage-Gate Approach is both a conceptual and an operational model for the development, marketing and removal of a product during its life cycle. The PLM- Process consists of seven stages. This structure is oriented on the life cycle of the product and has been validated in several projects by Detecon Inc. primarily in the Telco-Industry. Every stage consists of a distinct number of activities (organized in perspectives, refer to the next section), which have to be accomplished by specific process stakeholders at the given phase. The entrance to each stage is a gate; these gates control the process much like quality control checkpoints. Each gate is characterized by a set of deliverables as input, a set of exit criteria and an output. Gates are manned by senior managers that form the so-called Product Board. The Product Board acts as a gate-keeper that evaluates the results from one stage by a given set of criteria. Based on this evaluation, the product board can either decide whether the product idea proceeds to the next stage, re-starts at the previous stage or is archived.

Principle Two: Multi- Perspective

Product Lifecycle Management has always to cope with the conflict of objectives between the product marketing and the technical side. Several case studies have documented that this conflict often results in product failure in the market. Either because the product specification is too much technical driven and therefore far from the market demand or the product marketing has defined product specifications without collaborating with the technical department about technical feasibility beforehand. For solving that conflict of objectives three perspectives on the process have to be taken into account in order to ensure an efficient coordination and collaboration of the relevant departments or business partners.

In the market perspective all activities are assigned to that relate to the product marketing. Referring to the service industry, typical tasks like the definition of product specifications and the management of the product on the market-place belong into this category. By contrast the technical perspective subsumes all technical or production- oriented activities. Additionally the financial implications of moving the product through the life cycle are grouped in the financial perspective.

Principle Three: Metrics based Management

In addition to the process definition, the organisational and business components must not be neglected. Since products, and consequently the PLM-process, become more complex and involve internal as well as external partners along the value chain, there is a greater need to balance top- management control with the empowerment of self-managed, cross functional teams. As a prerequisite for achieving this balance, the company has to implement a metrics-based management approach in which teams are measured on strategic performance indicators such as development cost, time to market and customer satisfaction.

The definition and selection of the indicators is critical for the successful implementation of the PLM-process. Our developed and tested metrics-based framework can be used to systematically make business decisions about implementing PLM and maximizing your return on investment. PLM-process metrics based framework includes 5 dimensions: process performance, process quality, product quality, order and request quality and customer satisfaction. Implementation of such a framework enables a company to establish success metrics that they can use to measure the effectiveness of your implemented PLM-process and solutions.

By setting the weights properly, the development teams will self-steer to the greatest short- and long-term profit, which results in less coordination effort and efficiency gains especially for cross-departmental teams.

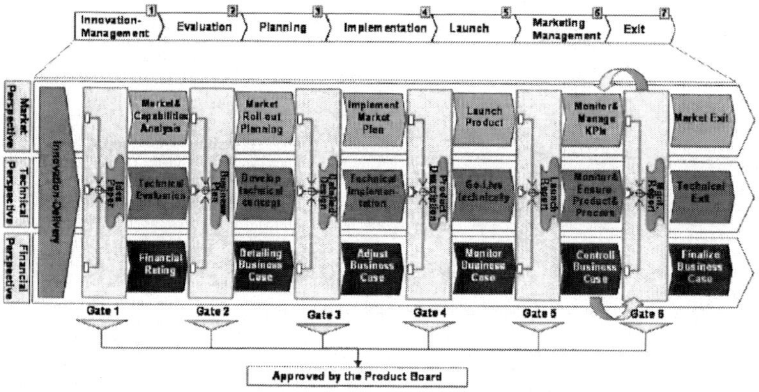

Figure 6. PLM-Process based on three principles

1.4 Lifecycle Value Management

Life cycle oriented product planning is a necessity for companies on the competitive market and requires a robust and systematic process combined with the right organizational environment. Lifecycle Value Management is the management of the product post-launch through the growth, maturity, and decline stages of the product life cycle. In the overall value chain, life cycle value management ensures that the extended products that are developed and implemented fulfil the market requirements and reflect the market demand. Lifecycle Value Management requires permanent attendance for analysing and planning for ensuring the customer benefits and the alignment to all company functions.

There are three different but supplementary building blocks of Lifecycle Value Management in the service industry:

Value oriented portfolio management process and IT-supported portfolio database

Strategic alignment and value based product portfolio management enable an effective product lifecycle management. Focusing on clearly defined products makes a significant contribution to effective portfolio management. The products can be positioned without overlaps and in accordance with market demand. Sales and marketing departments benefit from the increased transparency of the product range and the improved comprehensibility for the customers whereby the approach to customers is simplified and shortens sales talks.

Increased complexity of product marketing without an appropriate and efficient data foundation results in a lack of business transparency and low synergies in the value network. The modular structure of the portfolio promotes the use of common parts in the production processes, increasing the ratio mass production to overall production. A standardized product data structure creates the prerequisites for the continuity of the processes from production to sales and marketing. The development process for market products becomes faster and cheaper. The higher level of process automation improves production quality and process cost efficiency (e.g., by savings in IT operating costs and headcount).

Customer Needs Management (CNM)

Customer Needs Management (CNM) is a discipline within PLM that ensures that customer needs or requirements are communicated to all product development stakeholders across the lifecycle. Customer Needs Management effectively decreases the potential of design flaws and delays associated with misinterpretation and discontinuous communication, while improving productivity, efficiency and quality at the same time (Golovatchev et al. 2008).

At every stage of product lifecycle it is vital to take the customer reflections into account to achieve new product success and thus sustainable competitiveness. The Customer Needs Management strikes a balance between evaluation, performance controlling and steering of product portfolio. Specific value measurement is vital to understand an impact of Customer Needs Management on the steering of Product Lifecycle.

Reporting and controlling process regarding the product lifecycle management process

The referencing of different divisions (sales, resources and production planning; cost/profit accounting, etc.) to a standard product definition lays the foundation for the application of important controlling instruments (e.g. product success accounting). The unambiguous correlation of the basic data to business management indicators along the lines of a standard product structure provides staff and management with precise and timely information about all of the critical success factors that help staff and management to make the right decisions at the right time.

Implementation of these three building blocks combined with a modular product data structure ensures the “state-of-the art” Lifecycle Value Management:

Individual product performance information available at real-time.

Product Manager retains product accountability throughout life cycle.

Strategy for product growth, maturity, and withdrawal stages is defined up front.

Product replacement strategy also considered.

Systematic (e.g. annual) review and clean-up/optimization of product portfolio.

Case Study

A renowned company in the telecommunications industry carried out an extensive restructuring program which would enable it to maintain its position in a deregulated market environment. The objective was on the one hand to convert the previously technical driven approach for the product design (i.e., their orientation towards technical performance features) to an approach focusing on the customers’ needs and requirements. On the other hand, the aim was to develop and implement the integrated management approach, Next Generation PLM.

In the initial situation the PLM and the platform was not "state-of-the-art"(e.g. no withdrawal phase, missing of decision gates, long "time-to-market" etc). A portfolio management process was not designed and implemented.

The current portfolio structure was oriented on the organizational or technical structure and not organized from the customer's point of view. The product portfolio was characterized by a large number of product variants and features. All these products needed to be handled individually from an IT management perspective. This broad variety of products needed to be realized and implemented within all operative processes, IT-applications and – systems as well as sales information tools. This led to an enormous complexity that impedes the maintenance of IT-landscape and the management and optimization of the processes. No integrated IT-solutions were available at company and at its affiliates.

During the project the integrated PLM approach valid for the company and its affiliates was developed. Implementation of Next Generation PLM at this company showed the valuable benefits for solid product development, marketing and strategy.

PLM Meta-Model

Easier know how exchange and using of the "same language" during product development as well as fast and efficient communication between international partner

Introduction of the harmonized product definition and product portfolio for all national and international affiliate companies

The product portfolios across all of the company were to be reduced by 50% and integrated into a modular structure.

New ways to re-use of materials

Adoption of the product data platform at all international subsidiaries

PLM Process and Structures

Acceleration time-to-market up to 25% by several product groups

Reduction of quantity energy and material used in product development and production up to 20%

Efficient cost savings along the PLM process by using standard support system and re-using of modules and components (process costs saving up to 170 m USD/year in the product realization phase)

Effective and similar procedure of innovation and market management projects execution

PLM IT-Architecture

An implemented shared platform for document and project management.

One physical server is used to support separate product lifecycle management processes in all divisions and subsidiaries

Lifecycle Value Management

Sound marketing strategy due to the early recognition of market needs, standardized information and environmental issues

More detailed input for controlling for exactly allocation of revenue and costs to products

Simpler allocation at cost centres and cost unit

Introduction of the harmonized product portfolio for all national and international affiliate companies

Conclusion

In this paper the authors introduced a holistic PLM approach consisting of a process model, product Meta model, IT-architecture and a lifecycle value management concept. By applying the model to a real world case, the potential of the approach could be demonstrated. Future research activities have to be undertaken for ensuring a better synchronization of the different life cycles across the value chain taking the four pillars of NG PLM into account. Therefore we will further elaborate the building blocks towards managing the challenge of providing an interoperable PLM approach.

References

Abramovici, M.; Sieg, O. C. (2002): Status and Development Trends of Product Lifecycle Management Systems. In: Proceeding of International Conference on Integrated Product and Process Development, Wroclaw, Poland, p. 2122.

Ausura, B; Deck, M (2007): The new product lifecycle management systems: what are these PLM systems? and how can they help your company do NPD better? Visions Magazine. Online available at http://www.pdma.org/visions/jan03/plm.html, last visit 26.01.2007.

Golovatchev, J. and Budde, O. (2007): Next generation PLM – an integrated approach to product development in the service industry, Product Lifecycle Management: Assessing Industry Relevance, PLM-SP3, Interscience Enterprise Ltd., pp. 555–564.

Golovatchev, J., Hong, M.C., Budde, O. and Hiob, R. (2008): Next Generation PLM – Strengthen Competitiveness in the Telco Business, An Introduction to the 4-Pillars Approach of Integrated Product Lifecycle Management, Detecon, Bonn.

Mateika, M. (2005): Unterstützung der lebenszyklusorientierten Produktplanung am Beispiel des Maschinen- und Anlagenbaus. Essen: Vulkan-Verl.

Porter, M. E. (1985): Competitive advantage: Free Press New York.
Saaksvuori, A.; Immonen, A. (2004): Product Lifecycle Management: Springer.

Stark, J. (2004): Product Lifecycle Management: 21st Century Paradigm for Product Realisation: Springer.

Teresko, J. (2004): Auto Report: Information Technology The PLM Revolution. In: Industry Week-CLEVELAND OHIO-, Vol. 253, Nr. 2, pp. 32–38.

Thoben, K. D.; Jagdev, H.; Eschenbaecher, J. (2001): Extended Products: Evolving Traditional Product Concepts. In: Proceedings of the 7th International Conference on Concurrent Enterprising: Engineering the

Knowledge Economy through Co-operation, Bremen, Germany, pp. 27–29.

Wöhler-Moorhoff, Falk; Dieter, Stephan; Schwill, Matthias (2004): Telco 2010. Telekommunikation im Wandel: Die Karten werden neu gemischt. Published by Detecon. Bonn.

Ranking Management of Technology Conferences

HARM-JAN STEENHUIS

Eastern Washington University, Department of Management, 668 N. Riverpoint Blvd., Suite A Spokane, WA 99292, USA

HSTEENHUIS@MAIL.EWU.EDU

ERIK J. DE BRUIJN

University of Twente, School of Management and Governance, PO Box 217 7500 AE, Enschede, The Netherlands

E.J.DEBRUIJN@UTWENTE.NL

Several publications have appeared in the field of Technology Management which rank technology management related journals. Several different ranking systems exist for journals For example perceived relevance and quality, citation based and based on author affiliation. Academics also publish at conferences but we have not come across publications that rank conferences. Conference rankings are more complicated than journal rankings. Journal rankings are primarily for publishing purposes. Conferences on the other hand are attended by people for various reasons. In this paper a technology management conference ranking is developed based upon author affiliation. Ranking based on an analysis of author affiliation assumes that one important motive for a participant is to attend high quality research presentations. With that assumption the use of the author affiliation approach is an appropriate choice. Based upon existing rankings of institutes which offer technology management programs a ranking list of affiliations is developed. Subsequently, we compare technology management related conferences such as several years of IAMOT, PICMET, IEEE and the Technology Management Division of the Academy of Management based on this ranking list of institutes. The results provide information for authors, that is, the information can be used to determine which technology management conferences are most beneficial to participate in.

Keywords: Conference rankings, rankings, technology and management.

Introduction

Most faculty members are required to publish research as part of their tasks. For example, faculty at AACSB accredited institutions is required to maintain intellectual qualifications through among other things journal publications. Colleges can determine their own definitions of what is acceptable according to the standards of their institute. At many institutes this has developed into categories of journals, such as an A-list, B-list, et cetera. These lists are often based upon published journal ranking lists and are extremely important not only in determining faculty qualifications but also for retention and promotion decisions. Although maybe less regarded, conference publications are another method of intellectual contributions that can be evaluated for performance. To the best of our knowledge, no rankings have been published on technology management conferences. This paper provides a contribution in this area. It is developed by first discussing different journal ranking systems. The type of ranking system provides insight into how conferences can be ranked. Then, conference rankings are discussed, including the selection of a method and what the result, i.e. the rankings, mean. This is followed by a discussion of university rankings which serve as a base for the author affiliation index method for conference rankings and subsequently the comparative findings for several technology management oriented conferences are provided and conclusions are drawn.

Ranking Journals

Prior to discussing conference rankings, this section will first look at journal rankings since these methods provide insight into how conferences can be ranked. Three different types of methods for determining journal rankings can be distinguished. Each of them has a different orientation which (can) lead(s) to different ranking results. These ranking systems are based upon: perceived relevance and quality, citation, and author affiliation.

Perceived relevance and quality

With the perceived relevance and quality method, a set of journals is identified and a group of people, for example members of a specific academic society, is identified as respondents. These respondents are then asked through surveys how the journals on the selected list score on relevance and how they score on quality. Examples of this type of rankings are provided in (Barman et al, 1991; Barman et al, 2001; Soteriou et al., 1999).
A subjective element in this type of ranking is the initial selection of the respondents. For example, if IAMOT members are surveyed about technology management oriented journals then this ranking might differ from Technology and Innovation Division members of the Academy of Management. This explains how journals can be ranked differently even though the same survey instrument is used for ranking purposes.

Citation based

The citation based method determines which journals have been the most influential in a discipline. With this method a set of journals (set A) is identified for ranking purposes and additionally a set of journals (set B, which can overlap with set A) is identified for citations analysis purposes. The articles in journal set B are analyzed to determine which journals from set A are referenced most often. This determines the ranking of the journals (in set A). Examples of this type of ranking are provided in (Goh et al., 1997; Linton and Thongpapanl, 2004; Linton, 2006).
A subjective element in this type of ranking is the initial selection of the journals used for citation analysis (set B). For example, authors that use mathematical modeling approaches are likely to cite articles that are also mathematical modeling oriented. Therefore, if more mathematical modeling oriented journals are in set B, then it is expected that mathematical oriented journals in set A will be referenced more often than non-mathematical oriented journals in set A and therefore will be ranked higher.

Author affiliation

The author affiliation method is based upon ranking journals by examining who publishes in these journals and the affiliation of these authors. An example of this type of ranking, for operations management, is provided in (Gorman and Kanet, 2005). Ball and Rigby (2006) provide a similar viewpoint for management of technology oriented journals although not developed as detailed as Gorman and Kanet (2005).

A subjective element in this type of ranking is the initial selection of author affiliations. In the author affiliation method a ranking of affiliations is assumed and the journal ranking is based upon the relative number of publications from these ranked institutions. Another issue with this method is that it assumes that good researchers are at good universities and better researchers are at better universities. Although there are exceptions to this rule, it is likely that on average, in market-driven situations, the better researchers tend to be in the better universities.

Conclusion

Journal rankings have been determined to aid, among other things, tenure and promotion decisions by providing a sense of 'quality' of publications. There are basically three different methods for ranking journals, each of them is based upon specific assumptions which influence the ranking outcome.

Ranking Conferences

Publications of journal rankings are fairly common but this is not the case for conference rankings. The problem with conference rankings is that conference attendance and conference publications serve more roles than a scientific contribution. For example, one motive to participate in a technology management conference can be to meet colleagues in the area of technology and/or innovation management. A motive can also be to learn about new developments in an area of interest, or to satisfy publishing requirements. Another motive can be that the specific conference theme is of specific interest. Getting feedback and/or validation for on-going research can still be another motivator. Lastly, in some instances, the specific location of the conference can be an incentive to participate. Because of this variety it is therefore much more complicated and difficult to develop a conference ranking than developing a journal ranking due to the different factors influencing the participation.

In this paper a ranking of conferences is developed based upon the assumption that participation is based upon scientific 'quality'. In other words, this assumption is similar to the system for ranking journals. Obviously, if this method is used for ranking conferences then there are similar concerns with regard to bias as when they are applied for journal rankings.

To the best of our knowledge, a ranking for technology management conferences has not yet been published. It is noted that conference rankings in other disciplines have been published, in particular in the area of computer science. An example is provided by Citeseer (http://citeseer.ist.psu.edu/impact.html.) This ranking is based upon the citation analysis method. Another ranking is provided by Computer Science Conference Ranking (http://www.cs-conference-ranking.org/). They use an estimated impact of conference that uses the following weights: 40% for citation of papers, 20% for quality of referees' reports, 20% for availability of resources for students, 10% for indexing and 10% for the percentage of conference papers that are accepted or appeared in reputable journals. This type of ranking combines some of the different types of motives for attending a conference, for example financial support for students and quality of papers. Much of the weight is contributed to the citation of papers similar to the citation method for journal rankings. A study by Mathis and Zech (1992) looked at regional economics conferences and ranked based upon two indicators. First, they looked at the proportion of papers that were subsequently published in journals and second, they looked at the number of economists from top-rated universities who participated. This latter aspect is similar to the author affiliation method.

In this paper, we argue to stay focused rather than to combine multiple aspects such as by the Computer Science Conference Ranking. Combining different aspects introduces the aspect of weight, which is difficult to establish in a reliable way because not all attendees attend a conference for the same reason. Our primary focus will be one of 'quality', and we accept the limited scope of this selection. The advantage is that the ranking will be less ambiguous. Its disadvantage is that this type of ranking will only be relevant to those attendees who are interested in the quality aspect of a conference. However, we assume that this is the majority of the attendants.

In this paper a choice is made to use the author affiliation method. Each method has pros and cons for ranking as described earlier. The advantage of the author affiliation method is that it gives an indication of who is attending this kind of conference. Author affiliations provide two indications for quality. First, with the underlying assumption that affiliation is related to quality research, the authors from these affiliations are expected to present high quality research. Second, it indicates that researchers from top-affiliates are interested in a particular conference which is, assuming that these researchers are mainly interested in high quality research from others as well, another indication that the conference is of high quality. Note that high quality is equated here with the quality of the papers. Third, a higher proportion of papers (and attendees) from ranked affiliations, may make it more attractive for others to participate in the conference as well.

Ranking Affiliations

In order to develop a ranking list based upon author affiliation, a ranking list of affiliations for management of technology is required. These rankings exist. For example the Financial Times publishes a ranking list of the top-100 full time global MBA programs. For the author affiliation index, such a ranking is not sufficient because a discipline specific ranking is required, i.e. for technology management programs. Business Week and U.S. News & World Report publish these types of rankings but don't recognize technology management as a separate category and their rankings only include U.S. institutions. The Times Higher Education Supplement (2006) (http://www.thes.co.uk) provides a ranking in technology but this is based upon technological performance such as through licensing activities and start-up companies, not technology management programs and this type of ranking is therefore not appropriate. Linton (2004) has specifically ranked business schools on management of technology. With limited alternative technology management program rankings available this source has been chosen for determining the author affiliation index. Linton provides a list of international business schools that have an active management of technology research capability. Linton (2004) used a series of 21 metrics to develop this ranking which included quantitative aspects such as number of active faculty, the number of publications researchers have in the base journals during the study period, total publication history, and management of technology publishing history. It also included qualitative aspects such as the frequency of citations. Table 1 provides an overview of the top ranked schools including the metric papers per researcher. Appendix A provides the complete ranking.

Table 1. Worldwide top-19 management of technology affiliations, adapted from Linton (2004)

Rank	Affiliation	Country	Papers/researcher
1	George Washington University, Department of Management Science	USA	8
2	Portland State University, College of Engineering	USA	6.75
3	INSEAD	France	5.5
4	University of New Mexico, Anderson School of Management	USA	5.33
5	Institute of Super Tecnico	Portugal	5.25
6	Fraunhofer Institute of Systems & Innovation Research	Germany	5
7	New Jersey Institute of Technology, School of Management	USA	4.5
8	Stevens Institute of Technology, Wesley J Howe School of Technology Management	USA	4.33
8	Case Western Reserve University, Weatherhead School of Management	USA	4.33
10	Rensselaer Polytechnic University, Lally School of Management & Technology	USA	4.25
11	Conservatoire Natl Artes & Metiers	France	4
11	University of Massachusetts, College of Management	USA	4
13	National Centre for Technology Management, Obafemi Awolowo University	Nigeria	3.8
14	Delft University, Faculty of Design, Engineering and Production	The Netherlands	3.75

14	Brunel University, School of Business and Management	UK	3.75
16	University of Brighton, Centre for Research Innovation Management	UK	3.67
16	University of Washington	USA	3.67
18	University of Manchester	UK	3.62
19	Georgia Tech, Technology Policy & Assessment Center	USA	3.6

One of the problems of using a list is that rankings are sensitive to the nature of evaluation criteria. This for example occurs when looking at the number of publications and the importance of a publication. Linton (2004: 419) points out those publications in the base journals that either are related weakly or are irrelevant to management of technology are given too great an importance if they are counted as evidence of management of technology research activity. Linton (2004) therefore conducted more detailed analysis which led to a ranking with several tiers. However, this improved ranking based on three tiers includes only US schools. It is based upon recent research activity of the active management of technology researchers in the most management of technology oriented business schools in the US (Linton, 2004: 426). Table 2 provides insight into these tiers.

Table 2. Worldwide top-19 management of technology affiliations, adapted from Linton (2004)

Tier	Affiliation
1	Georgia Tech, MIT, Rensselaer, Rutgers, Stanford
2	George Washington, Michigan State, NJIT, Portland State, Stevens Institute of Technology, University of New Mexico
3	Case Western, Drexel, Harvard, Michigan, North Carolina State, Rockefeller University, University of Minnesota, University of Pennsylvania, University of Washington

Although this more detailed look at affiliations seems sensible it will not be applied in this paper because it only involved a limited number of U.S.-based schools. For this paper we therefore selected the original list.

Findings

The author affiliation index for conference x is determined by Gorman and Kanet (2004) as following. Let n(i) is total number of authors for article I, A(i) is the number of authors for article I from the top university set, B(i) is the number of authors not from the top university set, M is the set of conference articles for conference x, then the author affiliation index for conference x is:

$$AAI(x) = \frac{\sum_{i \in M} A(i)/n(i)}{\sum_{i \in M} [A(i) + B(i)]/n(i)}$$

The author affiliation index is the ratio of authors from top-schools divided by the total number of authors. Earlier analysis (Steenhuis and de Bruijn, 2006) has shown that in some cases the number of authors that appear on a paper can be quite large. Furthermore, in many instances this involves authors from the same affiliation. To control for a potential bias from this aspect only the first author will be used. The author affiliation index is therefore adjusted to a first author affiliation index (FAAI) which is computed as follows:

$$FAAI(x) = \frac{\sum_{i \in M} A_i}{M}$$

Where FAAI(x) is the first author affiliation index for conference x. A_i is a first author from the top-100 affiliations and M is the total number of articles at conference i.

For the analysis, the FAAI index is calculated for several years of IAMOT and PICMET, both are US-based Technology Management oriented conferences although the former alternates its annual conferences between the US and an overseas location, AOM, the technology and innovation management division of the US based Academy of Management, and an IEEE conference.

Table 3. Conference Rankings Based upon FAAI

Rank	Affiliation	Location	First authors from ranked affiliations	Total papers	FAAI	Average FAAI
1	Academy of Management 2002	Denver, USA	46	114	0.404	0.402
	Academy of Management 2003	Seattle, USA	70	169	0.414	
	Academy of Management 2004	New Orleans, USA	74	186	0.398	
	Academy of Management 2005	Honolulu, USA	100	255	0.392	
2	PICMET 1997	Portland, USA	113	485	0.233	0.204
	PICMET 2001	Portland, USA	76	399	0.190	
	PICMET 2003	Portland, USA	64	328	0.195	
3	IEEE –IEMC 2004	Singapore	18	90	0.200	0.200
4	IAMOT 2001	Lausanne, Switzerland	52	259	0.201	0.146
	IAMOT 2003	Nancy, France	46	424	0.108	
	IAMOT 2004	Washington D.C., USA	53	412	0.129	

Conclusion

In this paper an effort has been made to provide a rank for technology management oriented conferences from a participant perspective. Ranking conferences is more complicated than ranking academic journals because multiple reasons exist why people want to participate in conference. The assumption was made that one main reason to attend a conference is because of the quality of the conference. Several methods exist to determine the quality for journals. One measure, the author affiliation index was deemed appropriate to use for conference comparisons.

The approach was limited to the first author. The affiliations used in the analysis were based upon an existing international ranking list of affiliations. The analyses show that the Academy of Management conferences have the highest proportion of first authors from ranked institutions. This is followed by PICMET, IEEE and IAMOT. Based upon these findings, it appears that the US-based conferences receive higher FAAI ratings than non-US-based conferences. This may be a reflection of the affiliation index and how it was determined. For future research it is recommended that more conferences are included, in particular IEEE conferences and to re-examine the assumptions on which Linton (2004) based his ranking of affiliations.

Lastly, in this research the orientation was that of the participant but another side to conference rankings is the conference organizers viewpoint. Similarly to participants, who may make a decision to participate in a conference for various reasons, which may include the quality of the conference, conference organizers may also have various reasons for organizing the conference. For example, the Academy of Management conference is known for its quality standard and it has a rigorous review process to maintain the quality of the conference. The IAMOT conference also has quality objectives but another important objective is to allow less experienced people to present and learn from the presentation and the feedback. The research may be in earlier stages and might be less well defined but the lessons for the participant may be very insightful and are considered important for IAMOT conferences. Therefore, although common motives can be found for conference organizers, the weight for each motive can be quite different. This can also explain why the FAAI based figures are lower for IAMOT than for example for the Academy of Management conference.

References

Ball, D.F. & J. Rigby (2006) Disseminating research in management of technology: journals and authors, *R&D Management,* 36(2), 205-215.

Barman, S., R.J. Tersine & M.R. Buckley (1991) An empirical assessment of the perceived relevance an quality of POM-related journals by academicians, *Journal of Operations Management*, 10(2), 194-212.

Barman, S., M.D. Hanna & R.L. LaForge (2001) Perceived relevance and quality of POM journals: a decade later, *Journal of Operations Management*, 19, 367-385.

Goh, C.H., C.W. Holsapple, L.E. Johnson & J.R. Tanner (1997) Evaluating and classifying POM journals, *Journal of Operations Management*, 15, 1997, 123-138.

Gorman, M.F. & J.J. Kanet (2005) Evaluating operations management-related journals via the author affiliation index, *Manufacturing & Service Operations Management*, 7(1), 3-19.

Linton, J. (2004), Perspective: ranking business schools on the management of technology, *The Journal of Product Innovation Management,* 21, 416-430.

Linton, J. (2006), Ranking of technology and innovation management journals, *Technovation,* 26, 285-287.

Linton, J.D. & N.T. Thongpapanl (2004) Perspective: Ranking the technology innovation management journals, *The Journal of Product Innovation Management,* Vol. 21, 123-139.

Mathis, E.J. & C.E. Zech (1992) The quality of regional economics conferences, *Eastern Economic Journal*, 18(2), 221-228.

Soteriou, A.C., G.C. Hadjinicola & K. Patsia (1999) Assessing production and operations management related journals: the European perspective, *Journal of Operations Mangement*, 17, 225-238.

Steenhuis, H.J. & E.J. de Bruijn (2006), Location analysis of contributors to technology management conferences. In *Proc. Second European Conference on Management of Technology, D.* Bennett, B. Clegg, A. Greasley and P. Albores (eds.), pp. 656-663. Birmingham: Aston Business School.

Vokurka, R.J. (1996) The relative importance of journals used in operations management research, A citation analysis, *Journal of Operations Management*, 14, 1996, 345-355.

Appendix A: Worldwide top-114 management of technology affiliations, adapted from Linton (2004)

Rank	Affiliation	Country	Papers/researcher
1	George Washington University, Department of Management Science	USA	8
2	Portland State University, College of Engineering	USA	6.75
3	INSEAD	France	5.5
4	University of New Mexico, Anderson School of Management	USA	5.33
5	Institute of Super Tecnico	Portugal	5.25
6	Fraunhofer Institute of Systems & Innovation Research	Germany	5
7	New Jersey Institute of Technology, School of Management	USA	4.5
8	Stevens Institute of Technology, Wesley J Howe School of Technology Management	USA	4.33
	Case Western Reserve University, Weatherhead School of Management	USA	4.33
10	Rensselaer Polytechnic University, Lally School of Management & Technology	USA	4.25
11	Conservatoire Natl Artes & Metiers	France	4
	University of Massachusetts, College of Management	USA	4
13	National Centre for Technology Management, Obafemi Awolowo University	Nigeria	3.8
14	Delft University, Faculty of Design, Engineering and Production	The Netherlands	3.75
	Brunel University, School of Business and Management	UK	3.75
16	University of Brighton, Centre for Research Innovation Management	UK	3.67
	University of Washington	USA	3.67
18	University of Manchester	UK	3.62
19	Georgia Tech, Technology Policy & Assessment Center	USA	3.6
20	Institute for Studies on Scientific Research and Documentation, Natnl Research Council of Italy	Italy	3.5
	Maastricht Economic Research Institute of Innovation and Technology	The Netherlands	3.5
	National Chiao Tung University, Institute Management Technology	Taiwan	3.5
	Temple University, Fox School of Business and Management	USA	3.5
	University of Illinois-Urbana-Champaign	USA	3.5
	University of Salford, Research Centre for Design and Manufacture	UK	3.5
	University of Sheffield, Sheffield University Management School	UK	3.5
	University of Warwick, Warwick Business School	UK	3.5
28	Rockefeller University	USA	3.33
	University of Aalborg, Department of Production	Denmark	3.33
	University of Michigan, Industry & Operating Engineering College of Engineering	USA	3.33

31	Victoria University, School of Business & Public Management	New Zealand	3.2
32	Aston University, Aston Business School	UK	3
	CUNY Bernard M. Baruch College, Department of Management	USA	3
	Ecole Polytechnique de Montreal	Canada	3
	Korea Adv Inst. Sci & Technology, Graduate School of Management	Korea	3
	Louisiana Tech University, College of Administration & Business	USA	3
	Michigan State University	USA	3
	Rutgers State University, Faculty of Management	USA	3
	University of Dayton, Research Institute	USA	3
	University of Nijmegen, Nijmegen School of Management	The Netherlands	3
	University of Oslo, Centre for Technology, Innovation and Culture	Norway	3
	University of Pennsylvania	USA	3
	University of Sydney	Australia	3
	University of Tennessee, Stokely Management Center	USA	3
45	Monash University, Syne Business School	Australia	2.86
46	Leiden University, Centre for Science & Technology Studies	The Netherlands	2.75
47	Drexel University, LeBow College of Business	USA	2.67
	Erasmus University, Rotterdam School of Management	The Netherlands	2.67
	University of Oviedo, Department of Administracion Empresas & Contabilidad	Spain	2.67
50	McMaster University, MGD School of Business	Canada	2.63
	Tokyo Institute of Technology, Department of Industrial Engineering and Management	Japan	2.63
52	Chalmers University of Technology, Institute of Management Innovation & Technology	Sweden	2.6
	UMIST, Manchester School of Management	UK	2.6
	University of Cambridge, Centre for Technology Management, Manufacturing and Management	UK	2.6
	University of Strathclyde, Faculty Engineering	UK	2.6
56	University of Sussex	UK	2.54
57	Bocconi University, CESPRI	Italy	2.5
	City University Hong Kong, Department of Management	China	2.5
	Cranfield University	UK	2.5
	IMD, Lausanne	Switzerland	2.5
	Indiana University, Kelley School of Business	USA	2.5
	Lehigh University	USA	2.5
	Nanyang Technological University, Nanyang Business School	Singapore	2.5
	National University Singapore, Sch Business, Ctr Management Innovat & Technopreneurship	Singapore	2.5
	Oxford University	UK	2.5
	State University New York – Buffalo	USA	2.5
	Syracuse University, School of Management, Innovation Management Program	USA	2.5
	Tel Aviv University	Israel	2.5

	University Beira Interior, Faculty Engineering	Portugal	2.5
	University Calgary, Faculty Management	Canada	2.5
	University of Edinburgh, Department of Business Studies	UK	2.5
	University Grenoble, Inst Econ & Polit Energie, Natl Sci Res Ctr	France	2.5
	University Keele, Department of Management	UK	2.5
	University of North Carolina Chapel Hill, Kenan Flagler Business School	USA	2.5
	University Paris	France	2.5
	Universite Pierre Mendes, INRA-SERD	France	2.5
	University Tsukuba, Res Ctr Univ Studies	Japan	2.5
	University of York	UK	2.5
	Wayne State University	USA	2.5
80	MIT, Sloan School of Management	USA	2.38
81	University College of London	UK	2.33
	University of Laval	Canada	2.33
	University of Minnesota, Carlson School of Management	USA	2.33
	University of Reading	UK	2.33
	University of Zaragoza, Centro Politecnico Superior de Ingenieros	Spain	2.33
86	Glasgow Caledonian University, Department of Consumer Studies	UK	2.25
	Harvard University, Graduate School of Business Administration	USA	2.25
	Stanford University, Department of Industrial Engineering & Engineering	USA	2.25
	University of Padua, Pipartimento Tecn & Gest Sistemi Ind	Italy	2.25
	University of Twente, Faculty of Technology & Management	The Netherlands	2.25
	University of West England, Bristol Business School	UK	2.25
	Wilfrid Laurier University	Canada	2.25
93	Eindhoven University of Technology	The Netherlands	2.17
94	Aarhus Sch Business, MAPP Ctr	Denmark	2
	Arizona State University, College of Business	USA	2
	Assumpt Univ, Grad Sch Business	Thailand	2
	Cardiff Univ., Cardiff Business Sch	UK	2
	Curtin Univ Technology, Sch Management	Australia	2
	Free University Amsterdam	The Netherlands	2
	Hebrew University Jerusalem, Sch Business Adm	Israel	2
	Helsinki University of Technology, Department of Industrial Engineering & Management	Finland	2
	Hong Kong University Sci & Technol	China	2
	IFF IFZ	Austria	2
	Natl Chunghsing University, Department of Business Administration	Taiwan	2
	North Carolina State University, College of Business Management	USA	2
	Southern Illinois University, School of Business, Department of Management & Mkt	USA	2
	Stockholm School of Economics	Sweden	2

	Universita della Basilicata	Italy	2
	University of Bologna, Department of Management	Italy	2
	University of California – Los Angeles, Anderson School	USA	2
	University of Denver, Daniels College of Business	USA	2
	University of Groningen, Fac Management & Org	The Netherlands	2
	University Louis Pasteur	France	2
	University of Newcastle upon Tyre	UK	2
	Universita di Pavia	Italy	2
	Universite du Quebec a Montreal, Department of Management and Technology	Canada	2
	University of Queensland	Australia	2
	University of Wisconsin – Milwaukee, School of Business Administration	USA	2
	Vlaamse Instelling Technol Onderzoek	Belgium	2

PART 5

" Management "

- *ranking management of technology conferences*
- *Information system for disruption management of the emirates flight operations*
- *Technology management : best practices of the south African automotive supplier industry*
- *assessing and improving project management information system in a multinational company*
- *management of technology supporting center for enhancing competitiveness of small and medium enterprise in Egypt*

AN EMPIRICAL STUDY OF INFORMATION SYSTEM FOR DISRUPTION MANAGEMENT

M.R. Abdi

Bradford University School of Management
Emm Lane, Bradford, West Yorkshire
BD9 4JL, U.K.
E-mail: r.abdi@bradford.ac.uk

S. Sharma
Member of the Emirates Group Project Controller
E-mail: ssharma@emirates.co

Abstract
This paper investigates communication process and business drivers required for flight operations and disruption management. The disruption management is studied to derive an operational strategy, in which input resources and transformation process are reconsidered in order to deliver effective output services to passengers. Effects of different types of flight disruptions are studied and the role of Network Control Centre (NCC) is identified through subsequent stages The flight disruption at Emirates Airlines are analyzed along with their impacts on airport, aircraft and passengers through a real case study. The paper is also intended to investigate the influencing factors on the strategic flight operations/disruption management using the PEST analysis. Accordingly, an Information Technology (IT) framework is developed to manage the operations disruptions as well as flight operations.

Keywords: Management Information System (MIS), flight operations, Disruption Management, Network Control Centre (NCC)

Introduction

The airline industry is undergoing a looming phase by poor profits and scarred from terrorist attacks. Companies are laying-off staffs and trimming their services in order to overcome the drop in the number of passengers and rising cost for securities. More and more airlines are forming the global alliances such as Oneworld, Star etc. in order to benefit from each other's resources, which include additional routes as well as code sharing agreements. However, Emirates Airlines based on Dubai continues to grow and perform exceptionally well in the recent years when some airlines have been struggling to survive. Emirates Airlines is not targeting the price sensitive passengers and hence it has no plans to enter low cost services. In contrast, the strategy is to set long- haul flights with some regional routes (to feed long-haul flights and maintain the presence in the region). In addition, the company is not considering entering airline alliances (other than code-share type agreements) because it would affect brand value and restrict flight operations. Besides, alliances favor airlines rather than passengers. Emirates is pretty high on price if compared to low cost airline such as EasyJet and JetStar. However, Emirates offers high quality services and can be easily compared with some of the best airlines in the world like British Airways (BA). Emirates world class service and strong network in the region create an unbeatable competitive advantage against any other airline in the region.

Flight operations include the aircraft schedule and rotations, manipulation and altering of the schedules, crew scheduling and management and flight statistical analysis. The unit cost of flights is mostly affected by flight operations as the other costs such as aircraft maintenance cost, crew salaries, fuel consumption cost, etc are relatively rigid to change. Flight operations management for an airline provides an integrated solution to plan, monitor and respond to real time events, analyze traffic statistics and improve operational performance (Airline Software, 2001). As the company plans to double their fleet with over 130 aircraft by 2010, the quantity of air traffic at the central hub Dubai will greatly increase. Looking at the daily hub operation, seamless and reliable communication between all relevant stakeholders in Dubai is a critical factor for its future success.

Airline service quality depends on their ability to meet the requirements of their planned schedules. However, various events ranging from severe weather to the failure of a crewmember to report for duty inhibit their ability to always satisfy their schedules. In some instances, as in the case of thunderstorms, only a single airport may be affected for a few hours. In others, a large weather system can disrupt airline schedules over thousands of square miles for several days. Disruptions may have minimal effect in some cases on the airline. In other cases, it can become severe, causing the airlines to delay, cancel or divert substantial number of flights and imposing substantial cost to them.

In simple words, disruption can be defined as *an act of delaying or interrupting the continuity* (Hyper Dictionary, 2003). However in operational terminology, Clausen, J. et al. (2001) defines disruptions as a situation during the operation's execution in which the deviation from plan is sufficiently large that the plan has to be changed substantially. Flight operations plans are published well in advance and hence any disruption could lead to severe impact on the regular operation of an airline. A disruption in flight operations takes place when the observed situation deviates from the planned situation and the deviation on operation is substantial. In order to continue operations, intervention is necessary to resolve the infeasibilities resulting from possible disruptions. Considering the fact that the Emirates Airlines has a massive growth plan in the near future, it has to be expected that disruptions in flights may increase and therefore a new communication process must be in place to deal with, not only the daily operations but more with major disruptions and urgent situations.

The increase of traffic in the future will require the most efficient communication infrastructure to provide the necessary quality of service even in the case of disruptions and irregularities. This requires another look and redefining the existing operational activities performed in flight operations. Very few attempts have been carried out to handle disruptions in flight operations. Operations research methods have proven track record of delivering high quality solutions to a vast range of planning problems within the airline industry, production management and logistic (ORMS Today, 2001). For example, researchers at the Department of Informatics and Mathematical Modeling at the Technical University of Denmark have been working on applying Operations Research (OR) techniques for disruption management. The plan produced by OR-based decision support is intended to be applied on the day of the disruption, it can be adjusted to take last minute changes into account, or it can produce alternative plans well ahead of potential problems (Clausen, J. et al. 2001).

The paper is intended to investigate the current processes to handle disruptions in flight operations. Flight operations are studied to discover proactive measures in order to avoid disruptions while running operational activities. In particular, an Information Technology (IT) strategy is developed to manage the operations disruptions as well as flight operations. The proposed IT strategy can assist managers to explore existing disruptions and build a knowledge-based approach on the basis of historical data for handling the upcoming disruptions. The IT-based analysis can facilitate managing the disruptions and might result an alternative solution, which can eventually reduce the disruptions and the cost incurred. The data used in the research has mainly been gathered from interviews and discussions with different stakeholders and from the existing documentation such as Emirates Flight Operations Manual (2003).

Flight Operations Strategic Management

Flight operations of an airline needs an integrated solution being able to plan, monitor and respond to real time events, analyze traffic statistics and improve operational performance (Airline Software, 2001). Due to this reason, flight operations can also be known as Integrated Operations Control (IOC). In general, the key components of flight operations can be classified in the following categories (Abdi and Sharma, 2007):

- *Scheduling:* Scheduling handles the aircraft rotation, manipulation and altering of the schedules, print flow maps and time tables, and generate reports.
- *Crew scheduling and Management*: This caters to the crew bidding within legality parameters. It also involves tracking of crew records, training, qualification etc.
- *Flight Statistics*: It helps to record and analyze the traffic statistics.

Figure 1 illustrates the overview of IOC conversion process where the inputs are transformed to the customers delight. The input resources consist of transformed and transforming resources which are treated, transformed or acted upon. The transformation or the conversion process consists of material processing, information processing and customer processing (Slack, et. al. 2001). Material processing focuses on the baggage, freight in order to change their location from the origin of the journey to the destination. Information processing, changes the possession of the data (for example the flight map is delivered to the flight crew) and processes the information in order to generate the reports. Customer processing transports the passengers to the destination. IOC plays the role of the transformer and takes care of all the processes within the conversion process to deliver the output.

The fundamentals of airline operations are not different from those of any other firms operating in an open market (Shavell, 2000). However, the airline operations are very different from other operations in terms of disruptive events. Manufacturing firms and some service providers when confronted with disruptive events are able to react to preserve the uninterrupted flow of product and services to their customers. On the other hand, the customers of an airline are either trying to board or are already on the airplanes when flights are cancelled, delayed or diverted. There is nothing an airline can do to avoid inconveniencing their customers.

The flight operations might have high/low *visibility* aspect within the same macro operations. Some of its activities are totally visible to its customer e.g. catering, crew service, on time departure, baggage handling etc. Other parts of flight operations have relatively little visibility. For example aircraft planning or flight dispatch has very little customer visibility. The high visibility operations require staff with good customer contact skills. In other area, IT can provide user-friendly, real-time solutions for the customers which can lower the cost of operations and enhance the airline productivity. For example, an IT solution which can provide an alternate flight route map in case of a disruption without going

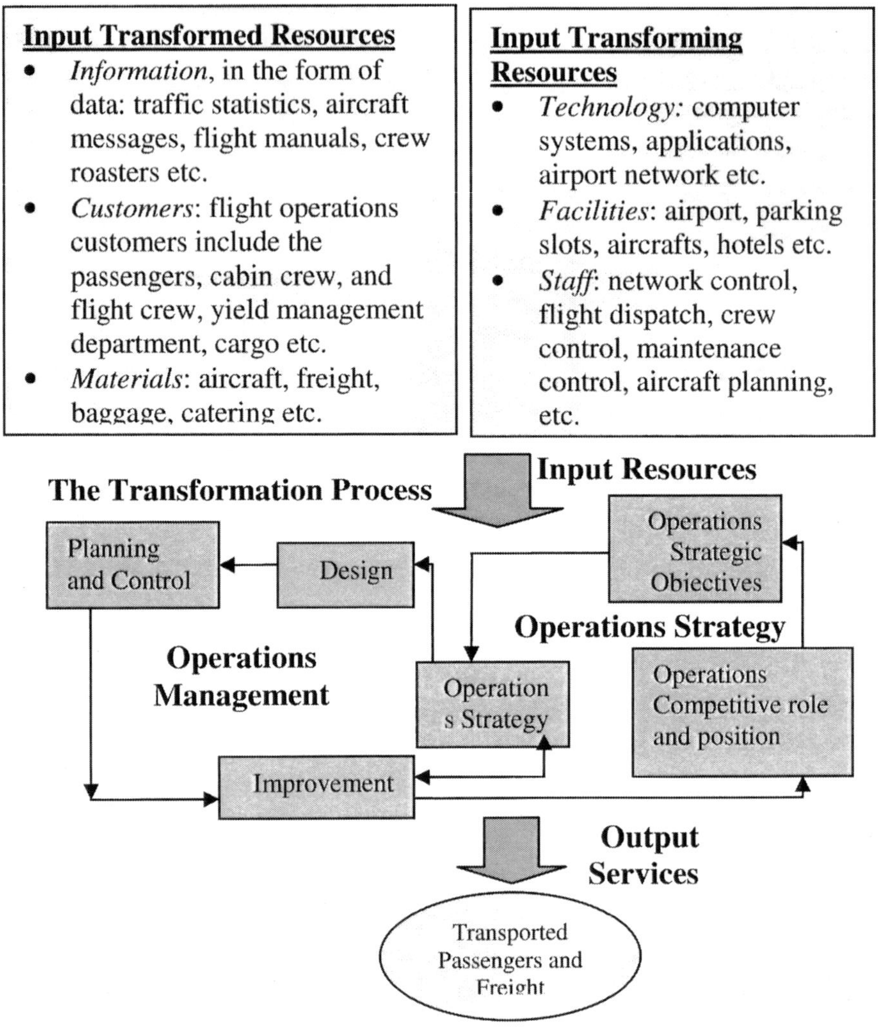

Figure 1. Flight operations management and operations strategy

into too many parameters or dimensions, will save the time and cost for the organization.

Emirates flight operations are high on *volume* dimension because of the high number of passengers transiting through the Dubai Airport. As per department of Civil Aviation – Dubai, a total of

18.06 million passed through the airport in 2003 (DIA - Statistics, 2004). There is the repeatability of the tasks which the staff performs in the flight operations. For example, a flight dispatcher repeats the same task for every flight which is dispatched from the Dubai airport. The staff follows the standard procedures set down in a manual, with instruction on how each part of the job should be carried out. IT can help in such a scenario by providing a business solution to achieve the standardization of the tasks.

Flight operations are relatively low in *variety* dimension because it operates according to a set schedule, published well in advance (e.g. the crew roster, aircraft plans are published three months in advance) and adhered to in a routine manner. Unless there is any disruption in the day-to-day running of the operation, the cost of operation is also low. The standardization and the regularity in the daily business of flight operations again highlights the scope of IT solutions in managing the operations.

As Dubai International airport is emerging as a transit hub for all the air travelers across the globe, it is relatively low in the *variation* dimension as well. The demands of Emirates flight operations is therefore relatively level. The crew, catering, flight plans and other activities of flight operations can be carried out in a routine and predictable manner which results in a high utilization of the resources. Figure 2 summarizes the typology of the flight operations.

Emirates IOC is characterized by high repeatability, well defined, routine and predictable operations which have relatively low unit cost. However, the disruptions in such operations escalate the unit cost tremendously. Thus one of the critical factors to the success of Emirates flight operations is to contain the cost of disruption handling. IT can play a major role in providing efficient solutions to disruption handling in operations. If Emirates is able to handle the disruptions in an efficient manner, it can not only improve its declining yield but it can also improve the company reputation by providing a quality service to their customer's delight.

Flight Disruption Management

Thunderstorms, equipment outages, fog and other unscheduled incidents, erratically occurring disruptions, inhibits air carriers' ability to fly published schedules, causing them to lose passenger revenues and incur additional costs. Because hubs only function effectively when passengers are able to make scheduled connections and often are located at congested airports, the effects of disruption are magnified when hubs are affected. These effects become most visible when the disruptions occur during peak arrival or departure. When faced with these severe disruptions, even if only one airport is affected directly, the effect ripples and relatively large number of flights may be delayed, cancelled or diverted. In simple words, Disruption can be defined as *an act of delaying or interrupting the continuity* (Hyper Dictionary, 2003). However in operational terminology, Clausen, J. et al. (2001) define disruptions as a situation during the operation's execution in which the deviation from plan is sufficiently large that the plan has to be changed substantially.

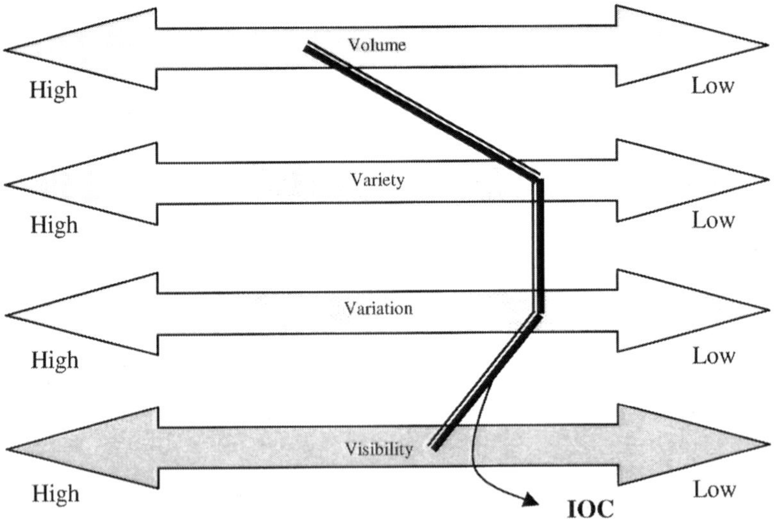

Figure 2. Typology of Emirates flight operations

Flight operations plans are published well in advance and hence any disruption could lead to severe impact on the regular operation of an airline. Emirates Airline operates on the *hub and the spoke* principle with Dubai acting as its single hub for the operations. The airline starts its operation from the hub and goes to the destination and returns back to the hub in a spoke fashion. The operations prime objective is to protect the hub from any disruption because the disruption in the hub can have a *Domino effect* which can last for several days. The effect will be more severe if there is any disruption in the *disruption window*. Disruption window represents the time when most flights land in Dubai from various destinations and depart to other destinations taking passengers from one part of the globe to another.

Delays, cancellations and diversions are the most visible evidence of the effects of these disruptions on the airlines. Generally, each of these results in aircraft and crews being out of position relative to planned itineraries. Passengers are inconvenienced as arrivals are delayed and scheduled connections missed. As a result, an airline may become responsible for the cost of alternate transportation, lodging, food and, if the delay is sufficiently long, a cash payment to compensate the traveler for any inconvenience.

As shown in Figure 3, airline business strategy, technology, staff and profitability are the most important elements for the establishment of an integrated communication for operations management along with disruption management. This Communication structure is a basis for a Network Control Centre (NCC), which can play a key role in the new communication system to handle the disruptions in an efficient manner. NCC provides a complete, fully integrated solution to control and manage all aspects of flight operations, from planning and development to collection of actual traffic statistics and analysis. In order to carry out normal operations, a plan is created and adjusted to take changing circumstances into account for the particular operations on the date. The plan is known as 'tracking process in operation management'. On the day of operation, the plan is implemented, and the operation is monitored during execution.

The monitoring and re-planning process is referred to as the control process. As opposed to the tracking phase, the time for re-planning in the control phase is so limited that the methods used for generating the original plan cannot be used.

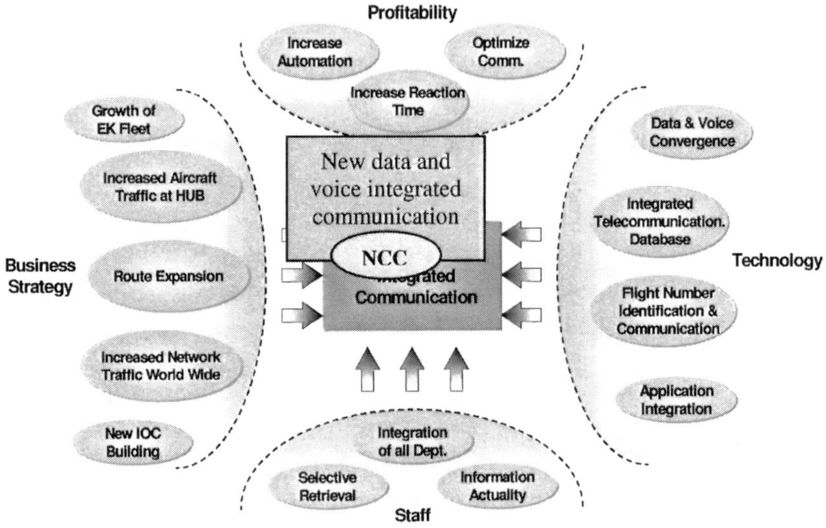

Figure 3. Communication process business drivers

PEST Analysis of the Flight Operations

The internal/external environments, in which flight operations are running, is very important to signify the operation management strategy. Accordingly, in the following section, a Political Economical Social Technological (PEST) analysis is presented to highlight the current position of the company from environmental perspective.

This section analyzes the external environment, which influences the Emirates and in particular Emirates Flight Operations in detail.

Each section will first discuss the factors at national level and subsequently the factors affecting the organization and the Flight Operations department will be discussed in detail. Considering the scope of the project, only factors which affect the disruption or disruption recovery plan will be discussed while discussing the factors affecting Flight Operations. PEST analysis will help to find out the Strength and Weakness in Flight Operations area which will eventually help in finding out the new IT solutions for the Flight Operations.

The PEST analysis is used to examine the Emirates flight operations and draw out factors critical to its success, given current and future macro-environmental influences. The PEST analysis will help to find out the strengths and weaknesses of flight operations identify the factors affecting flight operations. Accordingly, the external/internal environment factors, which influence the fight operations, are addressed. The influencing factors are discussed at national level, the organization level, and also within the flight operations department. Considering the scope of the project, among all identified factors, those which meaningfully affect the disruption or disruption recovery plan will be discussed.

Political/Legal Factor

Emirates Airline is highly sensitive to the government's action because of its close association with the Dubai government's policies. Emirates Airline plays a major role in helping Dubai government achieving its objective of promoting the city as a preferred tourist destination. As per Dasha, IT-Manager Emirates Airlines, understanding of corporate mission for the Emirates group can be summarized as "Support the development of Dubai in terms of tourists, transit passengers and cargo on a profitable base." The mission highlights the close proximity of Emirates Airlines with the national policy of Dubai.

Emirates Airline operates in a macro-economic environment characterized by stiff competition, stringent regulatory constraints governed by International Air Transport Association (IATA). In general, airline industry regulations are getting increasingly more complicated and require extensive admin efforts to maintain the compliance with those regulations. Some of those regulations were caused by 9/11 and other terrorist events and focused around strengthening immigration rules e.g. visa/travel authority requirements for Australia, US, UK, Malaysia.

There are many legal parameters governed by IATA which defines the way Emirates Flight Operation operates. Flight operation has to consider all these factors in disruptions. For example, a crew's maximum scheduled on duty time is fifteen hours on duty period containing International flying (Flight Operations Manual, 2003). Any interruption which causes the crew to stay for more than the maximum allowed time should prompt the Flight Operations to send the standby crew to release the crew on duty.

Other political factors which impact the flight operations are agreements like parking slots, landing/take off time restriction, agreement with other airport for standby landing etc. All these factors influence the disruption recovery plan. For example, if a flight is delayed, it might lose its parking slot and might end up paying more for the unscheduled landing at the destination.

Economic Factor

The macroeconomic environment which influences the Emirates Airlines comprises of two main components, the first one is Dubai, where the airline's hub resides and the second is airline industry, where Emirates operates.

Dubai's economy has undergone a rapid growth in last fourteen years and as per the market analysts, the United Arab Emirates (UAE) stock market is yet to pick up the steam and it is still undervalued (Cooper, 2004). As of 2003, The UAE's total GDP has grown by 6.2% to reach AED 73.3 billion ($20 billion) and the rapid development of the economy in recent years has been buoyed by the economy's diversification away from oil (Anderson, 2004). The GDP of Emirates has grown from AED 41 billion ($11.2 billion) in 1995 to about AED 98 billion ($26.7 billion) last year (Baik, 2005). The construction, Telecommunications, IT, Industrial, Tourism, and service sectors have all contributed significantly to Dubai's recognition as the regions trading and commercial hub. To summarize the Dubai economy, Dubai's economy is under the boom and likely to grow with the same rate for some more time.

Airline Industry on the other hand is racking up huge losses after the suicide attacks on the US. As per IATA, the airline industry group has lost $10bn within one year after the attack (Tan M., 2002). In addition to this, fuel price is also impacting the airline industry. As per Foss (2005), while improving economic conditions bolstered travel demands in the last three months of 2004, airlines seen losing more than $2Billion in fourth quarter due to fuel price.

The new regulations imposed by the European Union (EU) in the area of catering, crew health and safety, may cause changes to the existing business processes and working practices in order to achieve compliance. On overall yield will continue to decline across the airline industry, and hence efforts to further reduce unit cost may be required. These factors influence the microeconomic environment in which Emirates operates.

Disruptions in flight operations have a huge economical impact on an airline. For example, a delay in flight may lead to providing a hotel accommodation to the passenger, paying extra remuneration to crew, losing a passenger to a competitor airline, paying extra to the airport handling agency for the delayed take-off and the parking slot etc. etc. All these factors add up as an additional cost for the airline thus impacting the airline business.

Socio-cultural Factors

As per the Dubai Development and Investment Authority, the population of the UAE as of 2001 was estimated to be 3,290,000. The UAE is highly cosmopolitan environment and a large part of the population is non UAE national, primarily a mix of other Arab nationals, Asians and Europeans (Gulf News, 2002). Figure 4 presents the demographic details of UAE.

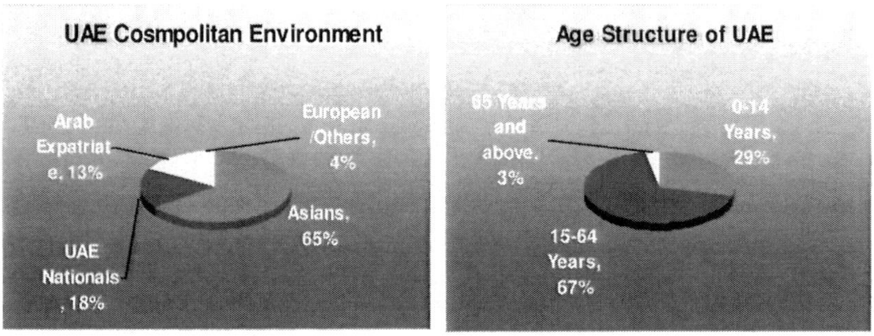

Figure 4. Demographic detail of UAE

Approximately 71% of the UAE population is male and 29% is female and the UAE population is expected to grow by 3.3% per annum to reach 4.15 million by 2010. The per capita income has remained at $20,000 over the past years for the UAE (Gulf News, 2002). Although Arabic is the national language of UAE but English language is the most important language for national, political and commercial communication.

Emirates Airlines also enjoy the benefits of a cosmopolitan culture and has a workforce from different countries. Being in the service industry which caters to the customers from all over the world, the cosmopolitan culture of Emirates Airlines proves to be a blessing to the business.

Culture plays an important role in the Flight Operation. The crew is supposed to be aware of the culture of the passengers and should provide the services to the customer's delight. For example, the crew in flight to Saudi should be very careful before offering liquor to the customers. Similarly the crew should not offend the Muslim passengers during the holy month of Ramadan by offering them snacks/drinks. Hence, during a interruption recovery exercise, it is important to place the right crew at the right place. The latest survey has revealed that Emirates is more successful internationally than regionally. It was expressed that Emirates is not perceived by local community as a national flag carrier of UAE but as international airline, and that affected their perception and preference for Emirates product. On a more objective note, it was noticed that Emirates product is not well tailored for Persian Gulf & other Middle East routes – in terms of cultural aspects and preferences of the passengers.

Technological Factors

Dubai is keeping up with the bleeding edge technology and providing the required infrastructure to its residents. Dubai Internet City, which is developed by Dubai government in order to promote e-commerce in Dubai, had a tremendous revenue growth in past. It was more than 100 percent during the first three years and this year alone the rise is expected to be more than 65 percent (Gulf News-b, 2004).

Emirates keep its fleet young and modern by investing in the latest aircrafts available in the market. Emirates Airline has placed the order for Airbus A380 which is the most technological advanced passenger aircraft in the world. Airbus states that A380 will use 20% less fuel and fly quieter, cheaper and more environmentally friendly than any other aircraft (Airbus, 2004).

Emirates Flight Operation uses the latest IT systems in order to handle its disruptions and the normal operation. Most of the applications are available to the users through Internet so that they can access it from any part of the world.

PEST analysis highlights the various factors or the external influences that affect the Flight Operations that are largely outside its direct control. The strength and weakness of the business environment helps the management to establish a strategic approach to their business. IT can proactively help in such a scenario by providing a complete solution to the management's strategic approach. For example, a stiff industry regulation indirectly increases the unit cost of operations and hence affects the business of an airline. The airline may take a strategic decision to reduce the unit cost of operation in order to bring the maximum profitability to their shareholders. In such a scenario, IT can automate the operational activities and reduce the unit cost by reducing the manpower required and the time required to achieve the same. For example, after 9/11 American government has asked all the airlines flying to the United States of America to submit the full detail of the crew flying to USA. It means a report has to be prepared before each flight which should have all the required information. Considering the Emirates mission of On-Time Departure, Emirates IT has provided a solution by automating the report generation process which not only saves time and manpower but it also provide an accurate human-error free solution.

Table 1 summarizes the Strengths and Weakness of Emirates Flight Operations business environment which has been derived from the PEST analysis discussed in the last section. These factors will be considered later while providing the IT solution for the flight operations.

Table 1. Strength & Weakness of Emirates Flight Operations Business Environment.

Type	Strength	Weakness
Political	Support of Dubai government.	Stiff airline industry regulations.
Political	Dubai's mission of promoting tourism in the region.	Interference of Dubai's government in Emirates policies.
		Stringent Immigration rules
		Agreement with other airports for the standby landing in case of disruptions.
		Agreement with other airport authority regarding the parking slots and landing rights.
Economical	Strong economy	Looming airline industry.
	Booming IT, Telecommunication, Industrial and tourism industry	Declining yield across the airline industry.
	Dubai's emergence as trading and commercial hub.	Additional cost involved in handling disruptions in Flight Operations e.g. hotel accommodation, parking charges, meal and other amenities voucher.
Social	Cosmopolitan environment	Being in service industry the crew needs to be aware of the culture and sentiments of their customers.
	High literacy rate and per capita income.	Challenge to provide maximum comfort to their customers during disruptions in flight operations like flight delay, cancellations etc. etc.
	Emirates workforce from different culture and nationality.	

Technological Technological	Dubai's support for the bleeding edge technology in terms of infrastructure.	Continuously changing technology requires continuous upgrade, training and development of the existing process.
	Emirates Airlines young and modern fleet.	IT solutions to provide 24x7 seamless access of the application to their users.

Flight Operation and Disruption Stages

In this section, different types of disruptions in flight operations are identified. The secondary data collected from the flight operations is analyzed in order to find out the factors which are critical for disruption handling. Emirates future growth is also considered while highlighting the impact of disruption in flight operations.

The NCC operational processes goes through different types in which operations are classified as per their severity. The NCC Operational processes in Emirates have been structured into four different stages where Stage 1 is classified as the least severe and the Stage 4 is classified as the most severe.

The four different types of NCC operational process as follows:

1) **Normal Operations:** Normal Operations are defined as simply normal day-to-day operations involving normal communication processes.
2) **Incidents:** An incident can be defined in many ways. Depending on the situation at the time. Normally an incident involves minor altercations to normal day-to-day operations, but could possibly increase its severity to a Disruption or Crisis.
3) **Disruptions:** A Disruption can be defined as a Major/Minor Disruption. This usually is entailed with weather conditions in Dubai such as fog. Fog is the leading cause of disruption in Dubai which forces the airport in

Dubai to close for a period until the fog clears, in turn all aircraft would be put into holding patterns or diverted to alternate airports. This is a main cause of concern and will force the Senior Manger Network Control (SMNC) and others to proactively tackle the situation.

4) **Crisis:** A crisis can be defined as simply an aircraft crash in either Dubai or at outstation.

Stage 1: Normal Operation

SMNC Conference Call

The NCC operates in 24x7 patterns, with 2-shift rotation (i.e. 12 Hours each shift). Each department within the NCC arrives at starts duty at different times. The SMNC is required to make a conference call at the beginning of each shift. The conference call is utilized to achieve a general overview (briefing) of the daily situation and to review any Maintenance, Airflow, Cargo, Ground Control and or any other general problems, which could be anticipated during their shift.

Maintenance Delay

A maintenance delay, which exceeds 24-hours, can always occur during normal operations. It can occur when navigational systems or technical problems arise which cannot be supported by the external airport, manpower and parts must be sent to external station or when an engineering problem arises and the aircraft needs to be checked by contract engineering departments at out station. The only difference between a maintenance delays at out stations and at the hub is that the NCC would more than likely initiate an Aircraft Change in Dubai which is shown in Use Case Aircraft Change.

Aircraft Change

An Aircraft Change could happen at any given time but mostly happens when delays occur at the hub, which exceed 24 – Hours. The SMNC business process is applied when an Aircraft Change is not avoidable. Sub-Process, which are the calling processes for,

"Structural Damage outside Stations" and "Maintenance Delay (Exceed 24 Hours)"?

Stage 2: Incident

May-Day Call Approach

A May Day or Pan call is not a mentioned topic in Category (A) or a Category (B) Crisis. Therefore, this scenario is worse case of an aircraft on approach into airport and has Engine failure and smoke or fire in the cockpit. The May-day approach can evoke three different processes based upon the severity of the damage. The use case describes all the processes in details.

Structural Damage

Structural damage can always happen but it is not often, although in the past 6 months there have been numerous incidents with lightning strikes (Neetan Chopra, 2003).

Stage 3: Disruptions

Major Disruption (Runway Closure)

The main cause for a Major Disruption in Dubai is fog. Fog is one the major problems in Dubai Airport and can cause a complete runway closure due to visibility. The use case covers the business process the SMNC uses when fog is not forecasted, and comes unexpectedly.

Stage 4: Crisis

Category (A) Crisis

Upon notification of an aircraft crash, hijack or missing aircraft, the SMNC would make the decision according to the EK ICC Manuel. A Category (A) Crisis is a worst case scenario up line or down line or in Dubai itself. The Category (A) Crisis means that many agencies need to be alerted and they need to react immediately.

Category (B) Crisis: Bomb Threat

Upon notification, a Category (B) Crisis can be seen and handled in many different ways. This use case deals with strictly worst case scenario in which a bomb threat call comes thought Emirates channels either in Dubai or at out stations. This constitutes that not all the same departments need to be contacted. A selection of departments would be informed. This scenario can also become a minor incident or a major accident according to the circumstances. There are two different types of bomb threat, a code - RED, and a code GREEN, which will be addressed in this use case.

Disruption Management via Information System

Decision Support System (DSS) can be defined as MIS at the organization's management level that combine data and sophisticated analytical models. It can also be defined as data analysis tools to support non routine decision making (Loudon, 2002). DSS helps managers make decisions that are unique, rapidly changing, and not easily specified in advance. They address problems where the procedure for arriving at a solution may not be fully predefined in advance (Robins, 2001).

Transaction Processing Systems (TPSs) are the basic business systems that serve the operational level of the organization. A TPS is a computerized system that performs and records the daily routine transactions necessary to conduct the business (Loudon, 2002). Figure 5 illustrates how the systems serving different levels

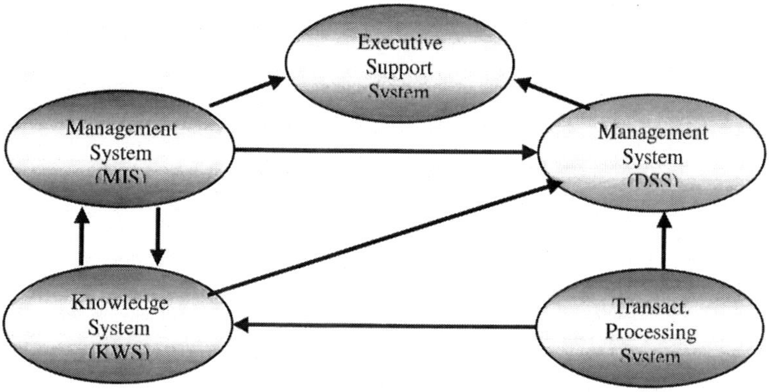

Figure 5. Interrelationships among Systems
(Louden & Louden, 2002)

in the organization are related to the other. TPS are typically a major source of data for other systems, whereas DSS are primarily a recipient of data from lower level systems. The figure also highlights that the DSS system is used by management which is a level higher than the user of TPS.

Disruption Management via Information System

In the new DSS solution, the existing TPS system of flight operations can feed the required data. Information about the crew legality and the current location can be taken from aircraft crew tracking system, the booked load on the incoming and outgoing flights can be taken from departure control and/or reservation system and the position of the aircraft can be obtained from operation resources control. All these systems are TPS systems, which will feed the data into the new DSS system. Any urgent message regarding any other constraint in the disruption can be obtained from inbound/outbound outstation messages/e-mail which are classified as Knowledge and Office Systems (KWS)). Hence, Emirates current TPS and KWS systems can feed all the available information about the flight operations into the new system.

Once the required data is obtained, the system can generate various recovery plans based upon sophisticated analytical theorems. These recovery plans can be presented in a user friendly format to the SMNC. The format outlines 'time' as the key factor to the success of the operations management because in case of disruption time is very critical. Therefore, the presentation of recovery plan should be self explanatory and should not demand any time from the control centre in order to understand the suggested method.

It is clear from the discussion that the NCC is managed by the SMNC. All other parts of IOC interact with SMNC and provide the required information to them. The input from various departments helps the SMNC to take decisions and initiate the required process. For example, in case there is an aircraft change due to a technical problem, Flight dispatch will report the issue to

SMNC which will take a decision and initiate the process by informing the Aircraft Planning. Hence all these departments work in close association with each other and help the SMNC to take a decision.

The base of the block represents the new IT system, which supports the entire NCC. The analytical theorem for the system is based on the customer values and cost-based decisions because the recovery plan should contain the cost of operations and it should also consider the customer value. Once the recovery plan is generated, the solution should alert the NCC with their roles and responsibilities. In order to alert all the stakeholders in the recovery exercise, the solution can be integrated with the existing Voice communication infrastructure in Flight Operations. The voice communication should cater to all the use-cases and inform the stakeholders accordingly.

The IT strategy should employ a combination of OR techniques – fuzzy logic, intelligent search method, classification of constraints and prioritization of objectives (Horner, 2004). All of the considerations of the recovery problem like lost revenue, passenger goodwill, cost of operation, crew maintenance etc. should be mathematically weighted in different combinations. The weights can then be adjusted as per SMNC's objective and business rules in order to generate the required recovery plan. The weight should be adjustable because it may vary depending upon the objective of SMNC. For example, during the Eid holidays or the festive season, Emirates priority will be to get the passengers home as early as possible (i.e. decision based on customer value). During other days they may want to minimize the cost of recovery (i.e. cost based decisions). Table 2 highlights the key stages and the corresponding activities, which should be generated by the new DSS system. The table also highlights its advantages and the respective stakeholders in NCC.

Table 2. The MIS strategic stages and its advantages

Stage	Activity	Advantages	Stakeholders
1	Control the order of flights arrives into the terminal.	Reduce taxi time, reduce ramp congestion and improve gate utilization.	SMNC, DOCS, Flight Dispatch, Yield Management Representative, Cargo Control.
2	Prioritize arrivals.	Permit smoother passenger and cargo/mail connections. Reduce miss-connections. Reduce disrupted passenger expenses. Reduce departure delay and the ripple effect of these delays and associated disruptions.	SKMNC, DOCS, Yield Management Representative, Cargo Control.
3	Assign aircraft and crew rotation	Automated check for crew legalities. Optimal utilization of seat factor. Reduce departure delay. Reduce ripple effect of disruption.	SMNC, Aircraft Planning, Crew Control, Flight Dispatch.
4	Identify flight for diversion.	Avoid diversion in situations where longer-haul or load-limited flights may not have sufficient fuel to absorb airborne holding delays. These flights will have landing priority over other flights, with higher fuel reserves.	SMNC, DOCS, Yield Management Representative, Cargo Control
5	Maintain bank integrity	Delay the early flights and speed up the late ones.	SMNC, DOCS, Flight Dispatch, Yield Management Representative, Cargo Control

Conclusions

In order to be competitive in the airline industry is to reduce the operation cost. It is important to find out the flight operation and how the possible disruptions can be handled. Accordingly, a flight operations strategy is presented to highlight input/output resources and the transformation process, which can facilitate the management and control of flight operations and disruptions. In order to cope with deal with disruptions, an integrated IT strategy is proposed based on business drivers and resources available. The disruption handling is considered in the existing communication process in the NCC, which can control the disruptions as well as operations. NCC is considered as the heart of the airline flight operations and the integration within the NCC is the key to the process success. The prime objective of the NCC is to ensure a smooth operation and to expedite the recovery of an irregular operation in order to minimize the adverse impact on revenue.

The current infrastructure at Emirates is facing enormous challenge in keeping up to the pace with current growth. The paper presents a new strategy and business architecture necessary to ensure operational reliability to support decision processes. Different flight operations and disruptions are classified with the corresponding stages at NCC to respond respectively. Any disruption in flight operations has a severe impact on passengers, airport and aircrafts. Disruption are analyzed and the impacts on airport, aircraft and passengers are discussed through a case study.

The proposed information system will ensure that everyone follows the same plan for the disruption recovery, which can be logged back to the system for the post-decision and learning exercise. However, the proposed IT solution should only support the decision of the SMNC and it should not take the decision for the NCC. In other words, the proposed IT solution should act as a decision supporter not as a decision maker. The DSS system is supposes to generate an optimal solution but it might not always be a practical solution. For example, if the system suggests transferring a passenger to a different carrier which is flying through a longer route, the passenger may decline to board the flight. In such a case, the solution provided by the DSS system will not be practical and the SMNC has to look for another alternative. For the disruption recovery exercise, it is important for the company to analyze the future state of flight operations. Future state analysis will ensure recommended initiatives are sustainable considering growth trend of the company. In other words, the analysis will make sure that the solution must contemplate tomorrow's problems, not just today's problems

References

Abdi, MR and Sharma S (2007), Strategic/tactical information management of flight operations in abnormal conditions through network control centre, International Journal of Information Management, 27(2), 119-138.

Airline Software (2001), Flight Operations Brochure [Internet]. Available from http://airline-software.com/Brochures/spectrum-flght-ops.htm [Accessed 02 January 2005].

Clausen, J, Hansen, J, Larsen, J, Larsen, A (2001), Disruption Management [Internet]. Available from http://www.lionhrtpub.com/orms/orms-10-01/clausen-hansen-larsen.html [Accessed 01 February 2005].

DIA – Statistics (2004), Dubai International Airport – Statistics [Internet]. Available from http://www.dubaiairport.com/DIAInternet/About+DIA/Statistics/ [Accessed 28 February 2005].

Flight Operations Manual (2003), Emirates Revision 18. General Civil Aviation Authority, United Arab Emirates.

Horner, P (2004), Disruption Management and Recovery [Internet]. Available from http://www.lionhrtpub.com/orms/orms-2-02/frcook.html [Accessed 11 February 2005].

Hyper Dictionary (2003), Disruption – Definition [Internet]. Available from http://www.hyperdictionary.com/dictionary/disruption [Accessed 01 February 2005].

Loudon, K and Loudon, J (2002), Management Information Systems. Seventh Edition. New Jersey, USA, Prentice Hall International. pp 43.

ORMS Today - Operation Research and Management Sciences (October 2001), Planning Problems. GA, USA, Lion Heart Publishing.

Reavy, J (2003), AGIFORS Airline Operations. Airline Update [Power Point Presentation] 18 June 2003, Auckland, pp 10 [Accessed 26 January 2005].

Shavell, ZA (2000), The Effect of Schedule Disruptions on the Economics of Airline Operations, [Internet]. Available from http://www.mitrecaasd.org/library/documents/shavell.pdf [Accessed 20 Jan 2004].

Slack, N, Chambers, S, Johnston R (2001), Operations Management. Third edition, England, Prentice Hall.

TECHNOLOGY MANAGEMENT: BEST PRACTISES OF THE SOUTH AFRICAN AUTOMOTIVE SUPPLIER INDUSTRY

MARTHINUS P FICK
Graduate School of Technology Management, University of Pretoria,
Pretoria 0001, South Africa
m.fick@esteq.com

ANDRÉ J BUYS
Graduate School of Technology Management, University of Pretoria
Pretoria 0001, South Africa
andre.buys@up.ac.za

There is tremendous pressure on automotive original equipment manufacturers to deliver innovative vehicles with high quality as efficiently as possible, which result in demanding requirements for its suppliers. Theory suggests that companies should adopt management of technology principles to align their business and technology strategies.
The study evaluated the competitiveness of a target population of South African automotive suppliers in the initial stages of their product lifecycle in relation to the global automotive supplier industry best practise. It is an exploratory study involving multiple case studies to determine whether South African automotive suppliers make technology investments based on pressures from the Automotive Original Equipment Manufacturers (OEMs), their multinational holding companies or their own business strategic alignment. The study contributed a benchmarking model of best practises focussed on collaboration, knowledge management, supplier integration, people and program management, IT infrastructure, requirements management, product design and manufacturing production. The benchmark determined which best practises the target population implemented and exposed a lack of technology strategies from a product centric point of view. The first-tier automotive suppliers involved in this study have very strong processes and practises backed by lots of experience and could most benefit from technology enablers for collaboration, knowledge management, people and program management, and supplier integration.

Keywords: Automotive Suppliers, Best Practises, Technology Management, Product lifecycle.

Introduction

Background

One of the industries that have been greatly affected by globalisation is the automotive and transportation industry. Consumers put tremendous pressure on vehicle manufacturers to provide innovative vehicles with high quality as efficiently as possible. AMR Research (2004:A2) indicated that the automotive industry's new vehicles development cycles changed from 4 years to less than 18 months, which place demanding requirements on suppliers. Globally competitive first-tier automotive suppliers need a systematic approach to improve business processes through the implementation of technology enablers. Porter (1988:211-212) identifies technology as one of the five forces that drive industry competition. There are a number of technology enablers that can assist automotive fist-tier suppliers in accomplishing these goals, but according to Walker (2000:16) the acquisition and management thereof must be executed to increase the competitiveness of the company, whilst balancing cost against benefit.

Current state of the South African automotive industry

As Africa's premier automotive supplier, South Africa's automotive industry is a source of national pride and sustenance. According to the NAAMSA Annual Report (2005:47), South Africa's broader automotive industry's overall contribution to GDP (Gross Domestic Product) is of the order of 7.2%. The automotive manufacturing industry is the leading manufacturing sector in the South African economy and the third largest contributor to GDP after mining and financial services. Automotive industry technologies are embraced in a broad range of partnerships with foreign companies.

To accommodate this and other high technologically advanced sectors, SA has one of the best infrastructure and service industries among developing countries. The South African automotive industry further receives support from government and other institutions to strengthen its global position.

Research problem

The primary purpose of the research was to evaluate South African automotive fist-tier suppliers against global industry best practices in terms of strategic alignment with technology investments that will enable them to compete in global markets. The supposition was that South African automotive suppliers make technology investments based on pressures from the Automotive Original Equipment Manufacturers (OEMs) or their multinational holding companies and not necessarily based on their own business strategic alignment. As stated by Pieterse (2005:7) the discipline of management of technology is still evolving. The research attempted to expose an area for study where the overall business strategy is defined by a multinational company although the South African supplier still need to maintain autonomy on defining their technology enablers to stay competitive.

Rationale for the study

In the mid-1990s South Africa shifted from import substituting industrialisation to trade liberalisation, a major drop in tariff protection and rapid integration into the world economy. This posed major challenges for the automotive component sector, which faced the need to become internationally competitive rapidly. Justin Barnes (1999) conducted a benchmark study on the competitiveness progress of the South African first-tier suppliers and followed it up with an analysis of the trends relating to firm level operational performance at automotive manufactures (Barnes 2000). Central to this was the use of a benchmarking model of key competitiveness drivers mostly in product manufacturing, supply and cost control (Barnes 1999:9).

Based on the assumption that a product's lifecycle can be defined as concept, development, manufacturing to end-of-life, then the second half of the product lifecycle was predominantly measured.

Manufacturers face an increasingly complex product innovation environment with faster product commoditisation, globalisation, outsourcing, and demand for more complex products, which are more keenly affected by time-to-market pressures. Yet their product development processes are frequently unable to simultaneously address key targets such as product launch date, product development cost, product cost, and product quality.

Research objectives and research questions

The main research objective was to determine which technology management best practises the South African automotive supplier industry implemented in the first stages of the product lifecycle and to determine the methodologies used to make technology investments. The research attempted to:

(i) Establish the current technology enablers used by automotive suppliers in the first stages of the product development lifecycle.
(ii) Position the South African automotive suppliers in relation to international best practises.
(iii) Identify the level at which automotive first-tier suppliers utilise current technology management theory in relation to business strategy.

This was accomplished through an exploratory study in the form of qualitative multiple case studies. Some of the key questions were:

(i) How does the automotive supplier industry make technology investment decisions and what are the motivating forces behind these decisions?
(ii) Which technology enablers did the South African automotive fist-tier suppliers adopt in relation to their global peer group?
(iii) Which best practices did the South African automotive fist-tier suppliers implement in relation to their global peer group?

Key attributes of technology management and best practices

This study attempts to give a picture of where the South African automotive supplier industry stands in relation to world standards and general best practice in the initial stages of the product lifecycle. From a system lifecycle engineering viewpoint the focus of the study is predominantly on the Need (Requirements), Conceptual-Preliminary Design, Detail Design and Development and the initial part of the Production phases as presented by Blanchard and Fabrycky (1997:19) in figure 1 below.

Khalil (2000), as quoted by Pieterse (2005:2), stated that a company defines its competitiveness by the "...ability to exploit ideas and resources in a timely, cost-effective manner in order to accomplish desired goals and objectives..." Furthermore, Khalil continues to warn that companies that are unable to harness and optimally utilise technology will lag and may not even survive. Automotive suppliers therefore need to invest in technology and implement it according to best practices for them to become globally competitive.

Proposed Model: Automotive Supplier Product Lifecycle Benchmark

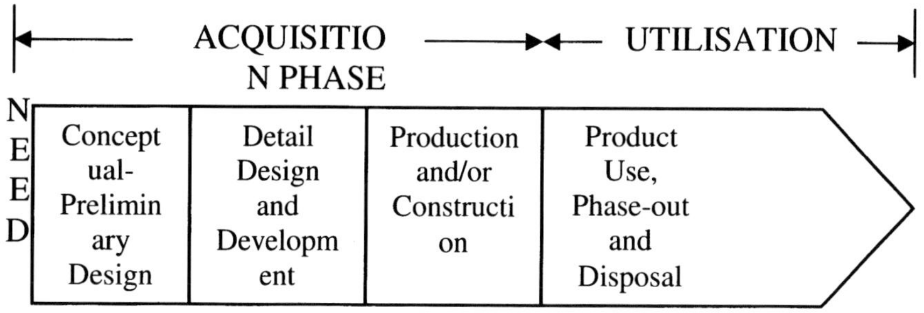

Figure 1. The product lifecycle Source: Blanchard et al (1997:19)

Product Lifecycles

The Automotive Consulting Group (ACG) (2003) has recently completed a study on product development in the supply chain. The purpose of the study was to identify and benchmark the best practices in product development so that automotive suppliers can compare themselves to others throughout the supplier hierarchy. Its core is an analysis of the various and sundry tools and practices employed by the suppliers for product development.

The Automotive OEM Lifecycle

Combining the results from the ACG (2003) study, the systems engineering viewpoint of Blanchard and Fabrycky (1997:19) and the product life-cycle management (PLM) footprint by Burkett et al. (2003:35) of the application that comprises a PLM system, the following generic lifecycle for Automotive OEMs evolved as shown in figure 2. It follows the approach to product design management known as quality functional deployment (QFD). Oakland (2000:39) describes QFD as a process for "...designing a product, based on customer demands, with the participation of members of all functions of the supplier organisation." The activities included in the QFD map of the automotive OEM lifecycle are shown in figure 2, starting with requirements capturing and planning to disposal.

The Automotive Supplier Product Lifecycle

In practice, the automotive supplier's product lifecycle is quite different from the OEM lifecycle, yet it is integrated and to a large extend driven by the OEM milestones. The automotive supplier's lifecycle delineates from the OEM lifecycle as presented in figure 3. The OEM product lifecycle starts with Requirements and Planning and continues with Concept Engineering. Suppliers do a small amount of this in Advanced Engineering. The next phase for the OEM is Sourcing. Automotive suppliers first have to get some business before they can source their tiered suppliers in the Business Planning and Proposal phase.

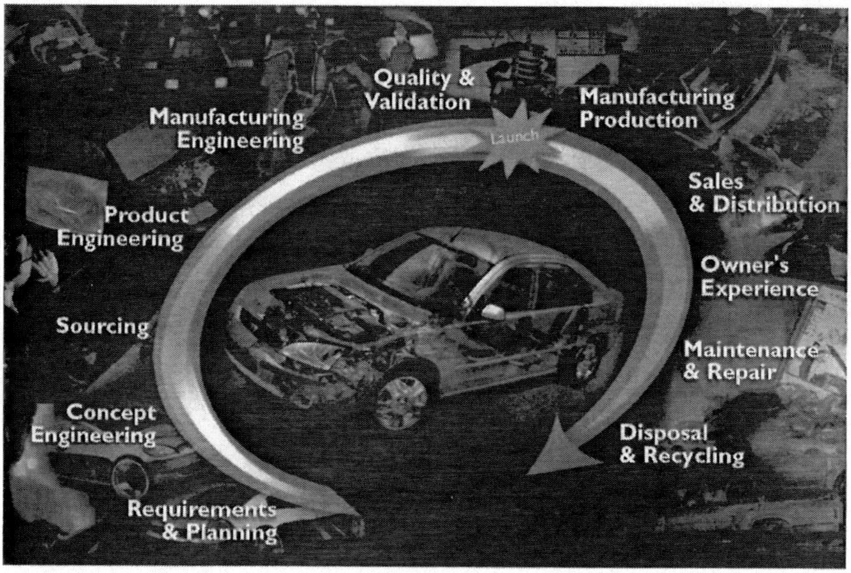

Figure 2. Automotive OEM Product Lifecycle

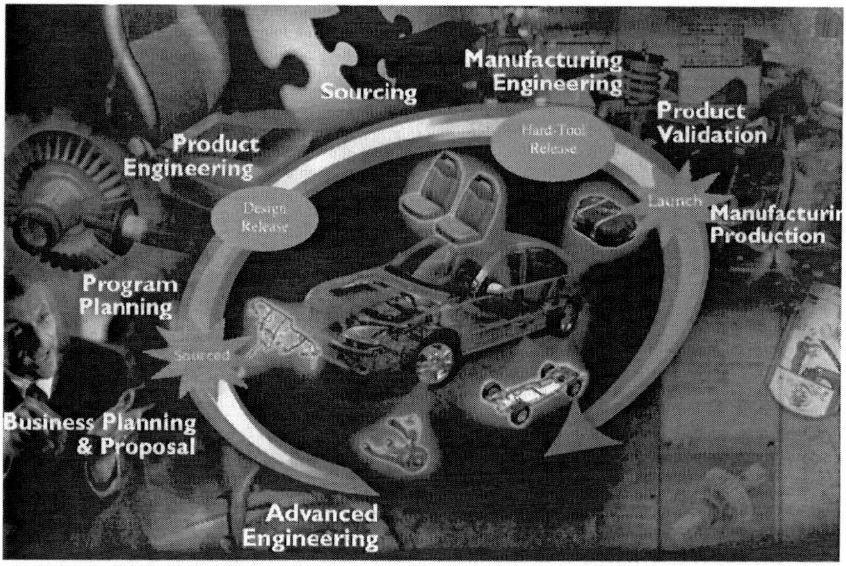

Figure 1. Automotive Supplier Product Lifecycle

As the OEM starts Product Engineering, the suppliers have to finish their Program Planning (resources, plants, people, etc.) before Product Engineering commences. For the most part, the next phase of their lifecycles is the same. The OEM does Manufacturing Engineering and Validation before they get to Launch. The supplier does those two things, mostly on tooling and plant, but they also have to deal with sourcing of their suppliers in the same period.

Critical milestones in the supplier lifecycle

A couple of significant milestones happen in the automotive supplier lifecycle between Sourced and Launch:

(i) Design release: Once a design is released, changes become more difficult since the same design affects all downstream suppliers and activities.

(ii) Hard Tool Release: Any changes following hard release not only impacts product engineering but also become extremely costly as existing tooling needs to be changed or new tooling made after this point.

(iii) Finally, the program goes to launch, the most important date on the calendar which the supplier may not miss at any expense.

Benchmarking model of activities, initiatives and best practises

A literature study on the key elements to measure best practises, supplied the basis to construct a model to measure the automotive industry's performance during their product development lifecycle. The model was constructed around core activities that should be conducted during the automotive supplier lifecycle to achieve business goals. The core activities broadly define the most important areas of best-in-class performance. They are further broken down into operational initiatives required to help respond to the business goals and finally the best practises that enables it. The benchmarking model in table 1 was constructed from selected components from the following sources:

(i) The 15 key elements by Marks (1995:127-148)
(ii) The results driven measurements by the Aberdeen Group (Brown 2005:5)
(iii) The Burkett et al, AMR Research (2003) on the typical tasks manufacturing companies struggle with.
(iv) The Automotive Consulting Group's (2003) indication of the elements that determine best-in-class practises for the automotive supplier industry, and
(v) The Board on Engineering Education of the American Society of Mechanical Engineers study by Evans et al (1996) with their twenty key elements of a product realisation process and the 56 best practises.

Research Methodology

The research was done by means of comparative benchmarking activities. These benchmarking activities were predominantly qualitative to establish the current best practise status and to find out what the target is, if any exist. The research was an exploratory study in the form of qualitative multiple case studies. The benchmarking activities were conducted through interviews with selected functional role-players within the automotive supplier industry. A questionnaire was drafted according to the benchmarking model presented in Table 1. International benchmarks were supplied by UGS, formally a business division of Electronic Data Systems (EDS), and represents data on the Standard Industrial Classification (SIC) codes 3711 and 3714 compiled by the Aberdeen Group (Brown 2005:5). The South African automotive suppliers were benchmarked according to this data as well as generalised best practises use.

Target population

The target population was the automotive suppliers in the South African industry. The sampling frame was the largest fist-tier suppliers as identified through three of the South African original equipment manufacturers based on total purchase value. Each of the OEM's in the local market is supplied by between 50 and 100 companies ranging from tier one to tier three. The top five suppliers (as indicated by the OEMs) from each of these groups were selected. The reason for selecting the top five was due to the digital maturity of the market where the top five suppliers of each of the OEM's were the most likely candidates where investments were made and measured.

Many of the smaller suppliers do not have any local research and development and would often supply components under license or standard bought-out components. In addition, there was some overlap regarding the OEMs these suppliers support. This led to eight companies on which the multiple case studies were performed. The selected automotive suppliers range from international subsidiary of multinational corporations to privately owned companies. Employee levels at these companies range form 416 to about 1500.

Qualitative methods to be used to capture analyse the data

The exploratory study was performed by means of structured interviews addressing best practices in the following key areas:

- Collaboration
- Manage Knowledge
- Supplier Integration
- People and Program Management
- Infrastructure
- Product Requirements
- Develop Product Design
- Manufacturing Planning
- Produce Product

The interviews were conducted according to the respondents' matrix of roles and activity mapping as depicted in table 2.

Table 1. Benchmarking activities, initiatives and Best Practices

Activities	Initiatives	Best Practices
Collaboration	Real Time Meetings/On-Line Access	Product Lifecycle Collaboration, E-Business Strategy
	Manage Workflow with Electronic Distribution of Data	Distribute Data, Notifications, Approvals
Manage Knowledge	Enterprise Information Management	Product Data Management, Data Management Scope
	Single Integrated/Synchronised BOM	Manage Product Configurations
	Lifecycle Configuration/Version Control	Configuration & Change, Manage Product Maturity
	Open Integrated Applications	Business Process & System Integration, Business Information Access , Management Systems Integration (PDM)
Supplier Integration	B2B Global Sourcing	Sourcing, RFP Communication with Suppliers
	Design Anywhere, Build Anywhere	Supplier Interaction, Supplier First Involvement, Supplier Uniformity
People and Program Management	Build/Manage Teams - Project Execution	Program Management, Team Structure, Program Roles, Staff Performance Goals & Measures, Empowerment, Product Lifecycle Process Definition, Manage Lifecycle Workflow, Product Lifecycle Process Utilisation, Continuous Improvement
	Knowledgeable Staff	Staff Training, Lessons Learned
Infrastructure	State of the Art Applications/Automation	Application/Automation
	Effective Info. Tech Support	Information Technology Support, Client/End User CPU, Server CPU Processor, Network
Product Requirements	Strategic Portfolio Planning	Platform Planning
	Requirements Driven Design	Requirements Planning, Benchmarking
	Systems Engineering Planning	Execution Planning
Develop Product Design	Digital MockUp & Virtual Prototyping/Simulation Driven	Digital Prototyping Process, Design Product, Check Interferences, Verify Product Performance, Test Prototypes
	Web Based - Multi CAD Visualisation & ViewMarkUp	Product Engineering, Team Involvement (manufacturing., etc.)
	Integrated Development Process with Reuse	Engineering Process Management, Standards, System Functional Reuse, Parts Management, Re-use Master Geometry Data
Manufacturing Planning	Integrated Manufacturing Planning	Process Planning, NC Programming
	Get Manufacturing Plan Right First Time	Simulation Validation, Test Manufacturing Equipment
	Manufacturing Resources At Right Time	Resource Management, Build Package Management
Produce Product	Lean Manufacturing	Shop Floor Information Access, Get Instructions to Manufacturing, Q/A Control Quality, As Built Verification

Table 2. Respondents Role and Activity Mapping

Role	Business Goals	Collaboration	Manage Knowledge	Supplier / Enterprise Integration	People / Program Management	Infrastructure	Product Requirements	Product Design	Manufacturing Planning	Produce Product
Area Sales & Key Account Managers	■						■			
Production and/or Engineering Managers	■		■	■	■					
Process Engineering Manager	■		■	■	■					
Product Development		■	■		■	■	■	■		
Manufacturing Manager		■	■					■	■	■
Quality Manager				■					■	■
Information Systems / Technology Manager		■	■			■				

The role mapping ensured that more than one respondent answered every best practise question during the one-on-one interviews. This enabled correlation of the respondents' answers within a specific automotive supplier. On average, eight persons mapped to the roles in table 3 from each of the eight companies involved in the study, were interviewed. This amounted to 63 interview schedules in which the activities were addressed. There are numerous operational functions within a company, which necessitates the use of this methodology to encompass as many dimensions of the company as possible.

Respondents' answers to the interview questions were recorded during the interviews. These answers were entered into an electronic database after the interviews. In situations where more than one person from a company supplied answers on the same activities, the respondents' answers were averaged to obtain a mean for the specific company. The companies' results where then combined according to their multinational status and mean results obtained from the respective answers. The results are shown graphically as 'spider charts'. The spider charts represent an averaged response to the best practise questions.

Results

The results of the data captured during the qualitative surveys at the eight automotive suppliers are represented as averaged response values to the best practise questions. Multiple respondents' answers on the same activity and at the same company were averaged to obtain the mean. The averaged results were then combined to obtain the total average for the target population. This is displayed in the form of spider charts (Figures 4 and 5). The spider charts display two averaged categories:

- Automotive suppliers that are international subsidiary to multinational corporations.
- Automotive suppliers who are privately owned local firms.

The core activities' spider graphs indicate the average of the respondents' answers on the best practices, current status and target relating to operational initiatives in their companies. (Figures 4 and 5) The results from the benchmark were compared to other suppliers in the global automotive industry to determine the current status of the companies involved in this study as segment of the South African automotive suppliers. In addition, they were mapped against industry standards or best-in-class performance. Both the privately owned companies and the international subsidiaries show close association to the current global industry standard. The captured data is displayed according to the core activities as set out in the benchmarking model. (Figures 4 and 5)

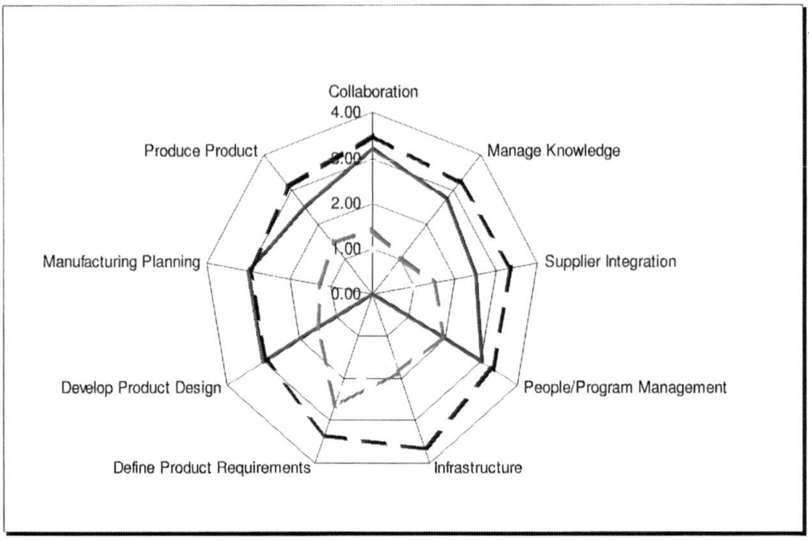

Figure 4. Overall Benchmark Maturity - Privately owned firms.

Table 3. Spider Chart Legend

Current	
Target	————
Current Industry Average	— — -
Target Industry Average	— — -

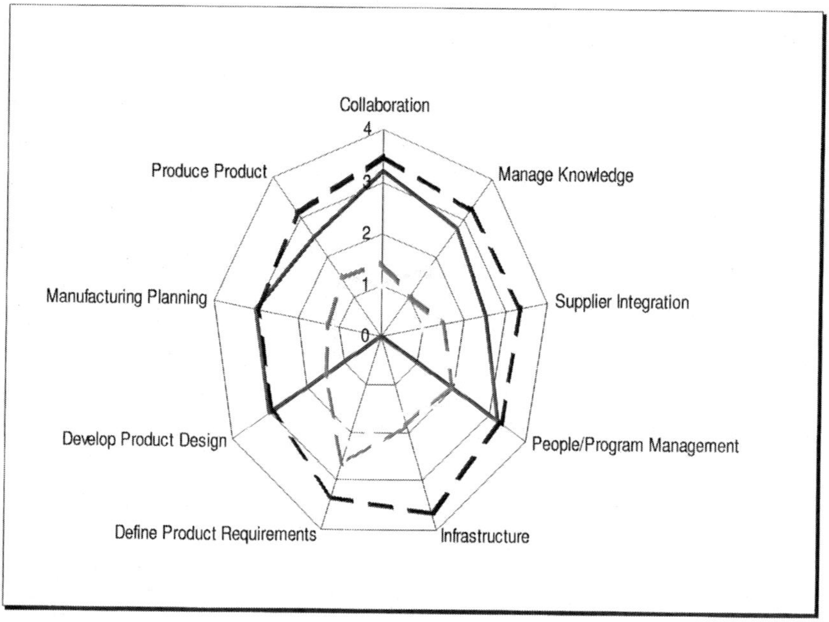

Figure 5. Overall Benchmark Maturity - Subsidiaries of multinational corporations.

Only two activities were affected by the influence of multinational corporations at three of the eight suppliers. These two activities are evident in figures 4 and 5 as Knowledge Management and People and Program Management.

The resulting comparison shows that the South African automotive suppliers in this study make technology investments based on pressures from the OEMs or their multinational corporation and not necessarily based on their own business strategic alignment. Furthermore, it highlighted the gaps that exist between global industry standards and the South African automotive suppliers.

Conclusions and Recommendations

The fist-tier automotive suppliers involved in this study compares very well to the global industry standard in best practises employed during the initial phases of their product lifecycle. The AMR Benchmark Analytix (2004:A10) shows that within the automotive supplier industry there is a prevalence of best practise use and e-business connectivity between automotive OEMs and the extended supplier enterprise, enabling some of the highest performance of all industries. In addition to this the AMR Benchmark Analytix (2004:A10) note that the automotive supplier industry uses very little application technology. In this light, it is not surprising to note that the fist-tier automotive suppliers involved in this study have the same tendency.

The fist-tier automotive suppliers involved in this study have very strong processes and practises backed by lots of experience. The four areas that could most benefit from technology enablers are Collaboration, Knowledge Management, People and Program Management and Supplier Integration.

The benchmark exposed a lack of technology strategies within the fist-tier automotive suppliers involved in this study from a product centric point of view in the pre-production phases of the product lifecycle. The industry could gain significantly by ensuring that the decisions taken in the pre-production phases, which could fix 70% of the production cost, are managed properly.

The benchmark also shows a gap between the business goals and the management of technology in that the industry is more reactive to technology than pro-active. Technology investment decisions are primarily taken based on existing vehicle programs or products that are produced. A technology strategy should map what the business want to achieve, how this will be enabled through technology and when will it be viable to implement the strategy.

It is important to note that it is typically not the same company, which is strong in all of the best practises as most companies make tradeoffs in their performance. Every company face a unique set of business goals, strategies, supply chain characteristics, execution capabilities and extend and blend of technology enablers they choose should be applicable to their unique situation … hence a technology strategy. More technology enablers are not necessary better, but rather how the company use it that differentiates them on their competitiveness.

References

AIDC Newsletter (2006). *Independent assessment measures AIDC's impact*, 17 September 2006, Automotive Industry Development Centre.

AMR Benchmark Analytix (2004). *Supply Chain Management Report Tier Automotive Industry.* AMR Research.

Automotive Consulting Group, Inc. (2003). *Best Practices in Product Development among Automotive Suppliers*, http://www.autoconsulting.com (visited on 28 July 2006).

Barnes, J (1999). *What changes operational competitiveness in the South African automotive components industry? Evidence from an international benchmark undertaking*, CSDS Research Report No. 20, Centre for Social and Development Studies, University of Natal.

Barnes, J (2000). *The competitiveness of the South African automotive components industry: findings from the 1999 follow-up study*, CSDS Research Report No. 27, Centre for Social and Development Studies, University of Natal.

Blanchard, BS and Fabrycky, WJ (1997). *Systems Engineering and Analysis.* Third edition. Prentice Hall.

Brown, J (2005). *New Product Development: Profiting from Innovation.* Aberdeen Group.

Burkett, M, O'Marah, K and Carrillo, L (2003). *CAD Versus ERP Versus PDM: How Best To Anchor a PLM Strategy?* AMR Research.

Evans, DG, MacKenzie, HR and Przirembel, C (1996). *Twenty key elements of a Product Realisation Process.* National Design Engineering Conference.

Khalil, TM (2000). *Management of Technology – The Key to competitiveness and wealth creation*. McGraw-Hill.

Marks, P and Riley, K (1995). *Aligning technology for best business results*. Santa Cruz, CA: Design Insight. pp. 127–148

National Association of Automobile Manufactures of South Africa Annual Report, 13th December 2005. http://www.naamsa.co.za/papers/ (visited on 30 Junie 2006)

Oakland, JS (2000). *Total quality management: text with cases*. 2nd ed. Butterworth-Heinemann.

Pieterse, E (2005). *The development of an internal technology strategy assessment framework with the services sector utilising total quality management principles*. University of Pretoria.

Porter, EM (1988). *The Technological Dimension of Competitive Strategy - Strategic management of Technology and Innovation*. 1st ed. Irwin. pp. 211–212.

Walker, M (2000). The importance of IT strategy. *Pulp and paper*. Vol. 74, Issue 9.

Assessing and Improving Project Management Information System in a Multi National Company

Omar Khalifa Gneiber

Associate Professor
Fo. Head of Industrial Engineering Department
Univerity of Garyounis
Faculty of Engineering
Ognieber@yahoo.com

Gebril Mohamed Zletni

MSc in Engineering Management
Associate Member of the Association for Project Management
gzletni@hotmail.com

Abstract

The overall aim of the research is to (re) design and prototype project management information systems to support engineering and project management services /operations for a multinational company (X). Four specific objectives have been derived from this aim and these are:

I. To analyze the existing facilities that contribute in information flow and identify constraints and problems;
II. To design a new/improved information system;
III. To test the new/improved system;
IV. To identify conditions for successful implementation of the system.

The analysis of the existing information system in the Multi National Company was done by visiting the study area during fieldwork. The aim was to identify the main stakeholders i.e., Clients, Consultants and Contractors, of the project information system, their information requirements and any bottlenecks existing in the system. Also analyzed, was the environment (both internal and external) in which the company operates by implementing SWOT analysis method.

It became evident that the current system is not effective. Equally, the company's Information System (IT) infrastructure is not up-to-date, and requires modernization to promote it up to the level of competitors. The system is slow and lacks resources both human and non human. An information system to help conceptualize the various components of project management and engineering services was therefore developed to serve as the basis for developing the database to store and manage relevant information. The functions and data requirements of the system developed are detailed.

The information system, Project Document Tracking Register Information Systems (PDTRIS) was implemented using Paradox database and Microsoft Excel as an export facility to enrich reporting variety, and this was tested for a few months to determine if it answered user information needs.

Certain conditions must be available for the system to be effectively implemented: skilled manpower, evidence of what the system is capable of doing, financial resources, specific responsibilities in maintenance of the system and extending the responsibilities / upgrading the capabilities of the custodian of the information (incoming and outgoing) i.e., a central document control office, are among the major conditions required.

1. Problem Statement

It should be noted that the exact name of the multinational company, the case study of this paper, and its shareholders could not be provided in this paper because they wanted to remain anonymous as some areas of project information systems failures were shareholders generated.

The current functions and structure of the multinational company, the case study of this paper, hereafter "Company" presents a highly centralized system with all authority vested in its senior management. The structure was set up like this so as to exercise control on all other headquarter departments and construction sites.

As a result, flow of information within the company takes long to process through a lot of junctions. The communication channel and information flow is static and rigid based on out dated procedures, which were set long time ago, since 1997 [1], where the internal and external challenges, and business environment were different compared with present time.

The process is entirely slow and routine which results delay in the processing and analysis of data because of large amount of data, which the company handles.

Information sharing among departments /sections and sites within the company is not easy. It is hampered by the bureaucracy of writing formal requests to the heads of departments seeking authority to access data required. The company services are overwhelmed by the contractor's submissions, the client's instructions and have to contain the situation with its meager resources. Resources (both human and non human) allocation to company situation is done and managed according to some fuzzy criteria without considering the overall picture of current and future project. Sometimes the management compelled to implement this policy to satisfy the Owners requirements.

In addition, the company's IT infrastructure and systems require modernization to bring them up to the level of the European peers and competitors [2]. The current IT system was developed on an ad hoc basis during sanctions and was not able to use the best available technology.

A considerable improvement in the IT infrastructure and systems is deemed to be business critical to improve company productivity. In particular remote sites must have fast effective communications to the head office and relevant contractors.

In order to successfully analyze information system within the multinational company, an integrated and holistic review of the existing situation is required. What is required is an improved project management information system.

2. Methodology and Approach

The structured system development methodology was adopted [3] and it takes an objective view of reality; provide elaborate planning guidelines, techniques and case tools. The method includes iterative activities which mean if there is a problem in one stage one can look at the earlier or later stages to rectify the problem. Figure (2.1) shows the main phases in system development using structured methodology. There are four major phases involved.

System strategy and planning, identifies and defines the problem and also how this problem may be solved. Goals and objectives are defined here and a plan of how to achieve them is outlined.

System analysis phase, which is, a detailed analysis of the existing information system is conducted. This phase defines why this information problem, where the problem is. It is in this phase that information requirements are found out and the constraints in meeting the information demand. System design focuses on the conceptual and logical design of information system looking at the function and data models. The physical design on the other hand looks at design in terms of hard and software databases and interfaces.

The implementation phase looks at building and testing the system; and integrating the separately developed systems among other objectives.

This methodology was used to develop an Information System for the Multi National Company, but with some modifications to suit the environment of the research, the time and resources available for the research.

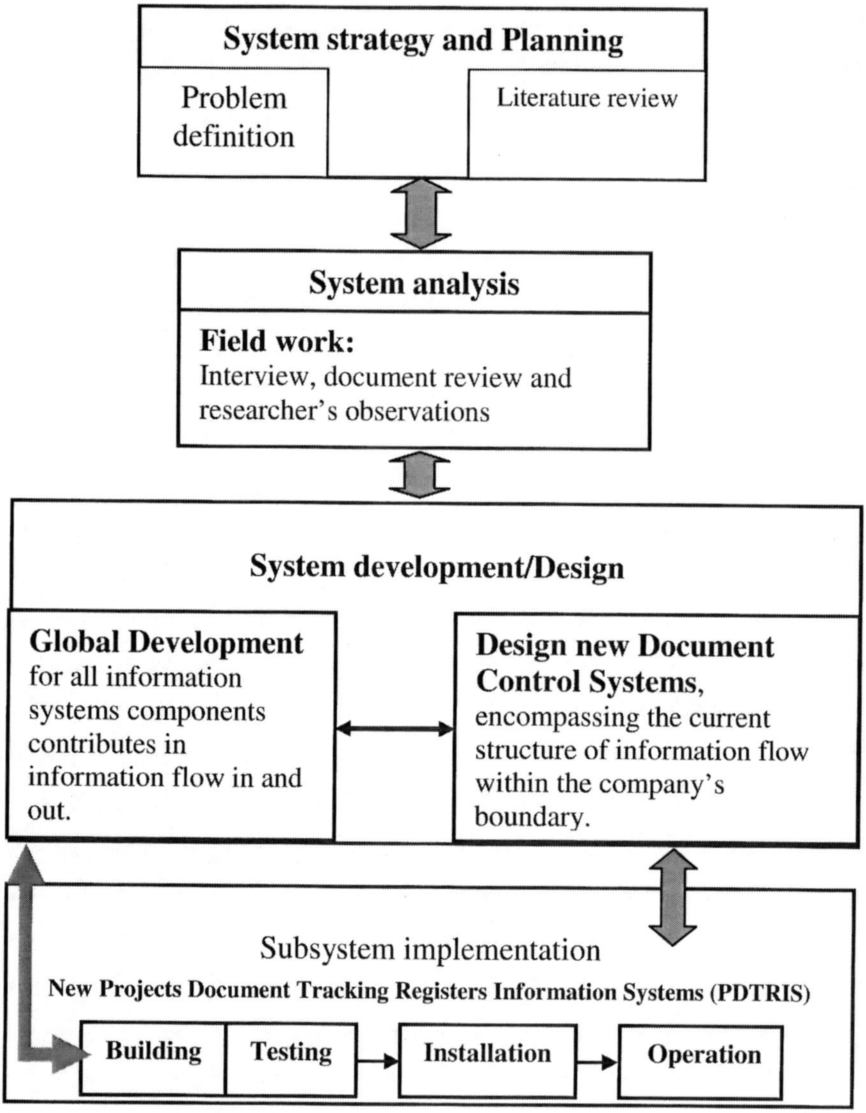

Figure (2.1) Structured System Development Methodologies, Steps and Interactions

3. The Analysis of the Existing Information Systems for the Multi National Company

3.1 Environmental Analysis

SWOT analysis was implemented briefly for analyzing internal and external environments in order to attain a systematic approach [4] and to assist the design stage and support the recommendations for further improvement. See figure (3.1).

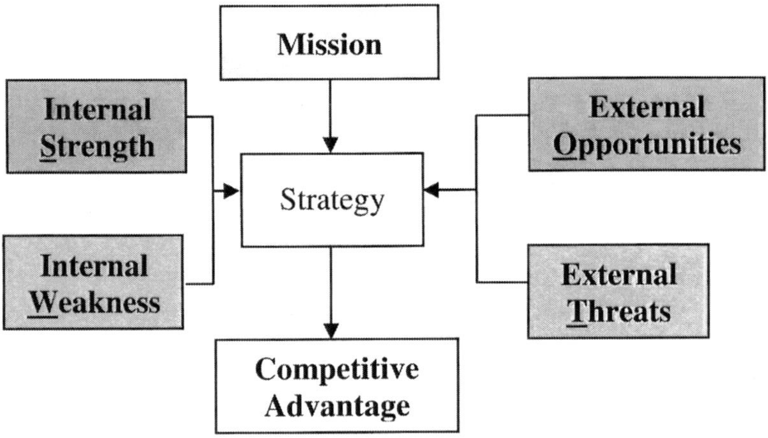

Figure (3.1) SWOT analysis [5]

Let us now discuss SWOT factors which were concluded from field work i.e., interviews and observations.

Strength

- The company is considered as the Market leader in engineering and Project Management services for large projects, water supply projects, and the Project Management Information Systems (PMISs) is already in place but it requires further improvement with the surrounded market, particularly at this opening stage.

- Skilled/experienced staff (manpower resource pool) inputs/outputs could improve further with the support of an enhanced information and communication technology (ICT).

Weakness

- Company's decision makers need to understand the local market in Libya. Translating market requirements into operation decisions is an essential factor. Business Development team to be established, at the headquarter office as an external eye on the surrounding environment.
- User resistance to change, particularly the document control staff. This has been noted during design stage at brain storming sessions of the new PDTRIS.

Opportunities

- The Company should start to establish chances of business opportunities with other organization in order to penetrate the local market and to start even with small project i.e. design package, feasibility studies, where this will build a name for the company and small business will bring large business.
- Improves document control process, which reduces administration costs. The intended PMISs is to move towards paperless office reducing administrative overheads.
- Improves communication and coordination between project.

Threats

- Other companies, competitors, have approached the Company's main clients and have already obtained chances of business, particularly large projects, their competitive advantages were low cost and lower standards.
- There is a possibility of financial losses if the implemented PMIS is mismanaged.

3.2 Outdated Project Tracking Registers

The existing document control system (tracking registers) is outdated, figure (3.2) and cannot cope with increased data flows, more frequent revision and difficulties in ensuring currency of data. The review revealed a number of problems and anomalies which have an impact on the company's (study area) preferred ways of working.

- Working under DOS environment that is no longer used/ supported (Data Ease system).
- Rigid system difficult to modify or accept new requirements and data entries either data entries screen or reports screens. See figure (3.3).
- Centralized and inflexible system, data controlled only by the document controllers only staff, restricted access to others. Not network utilization.
- Not user friendly and not compatible with other window-operating environment.
- Impractical data entry system as it was designed for major project, and to satisfy very tight contractual requirements. see figure (3.4).
- Designed to serve data entries of one project at a time, which has an impact on the daily process in terms of cost and time i.e. dedicated document controller.
- Limited production of status reports.
- Length of time it takes to reinterpret a data set after new data is collected (duplication of data).

The system was designed in 1990, for a strategic project, an estimated budget of circa $ 8 billion, to satisfy the flow of documentation as stipulated in the contract, Consultant vs. main contractors.

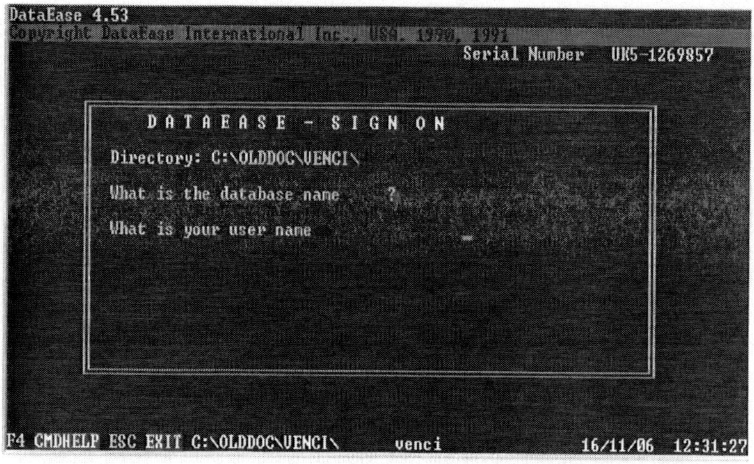

Figure (3.2) Main Menu of the Existing Document Tracking

Most of the options are not utilized by the document controllers as well as not working as the system failed partially.

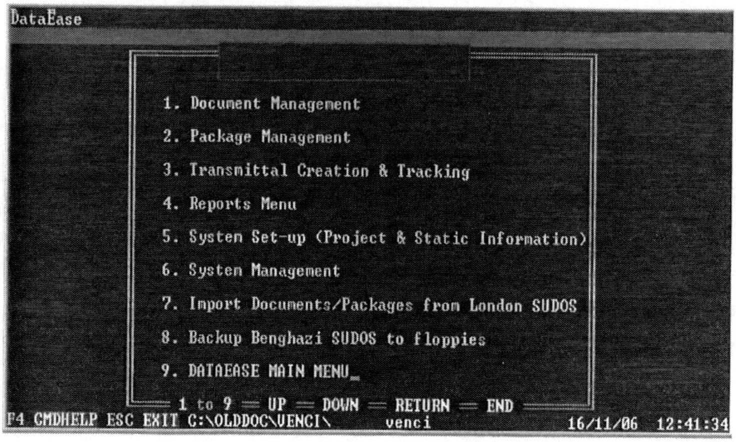

Figure (3.3) DTR's option Selections

Designed to reflect engineering process based on different organization and conditions which were changed since 1996. Assigned fields for data entries do not reflect the existing requirements either contractually or technically. Review process for project teams has been changed *dramatically.*

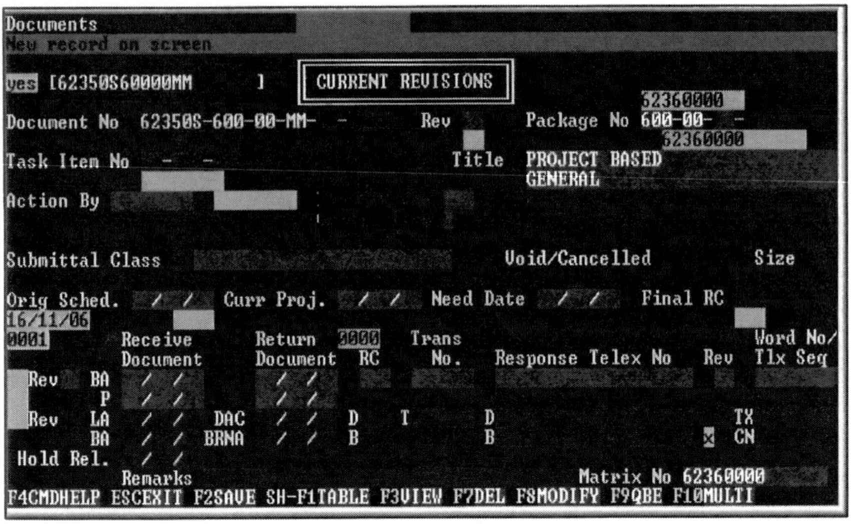

Figure (3.4) DTR's Data Entries Menu, New Transmittal /Package/Documents

3.3 Lack of Comprehensive Engineering Statistics Analysis

The major problem is lack of comprehensive reporting analysis within the existing tracking registers (existing document tracking registers). Incoming and outgoing data are in the tracking system but are not supported by reporting system. Various reports such as overdue reports and status reports and summary reports are not considered in the system, as a result overdue causes and trend are not understood resulting into poor decision on what corrective actions and prevention measures to be implemented by top management.

4. Design of Document Tracking Register System (PDTRIS)

A Project Tracking Register System (PTRS) defines the processes, the data flows involved projects information to be effectively analyzed. The processes are defined first at context level which gives a global overview of the system and secondly at top level which gives a detailed view of the system. The system being

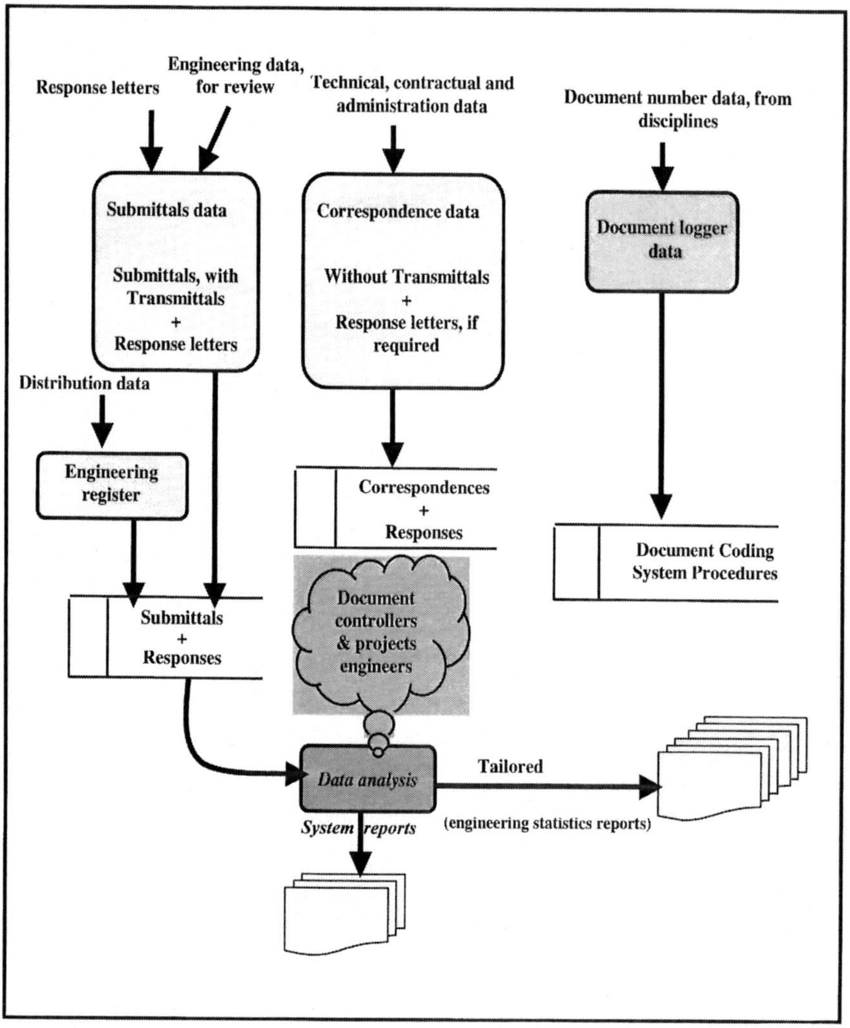

Figure (4.1) PDTRIS Top Level DFD

developed is mainly to satisfy and to respond to the current information requirements of the users while looking at what it can perform in future. Figure (4.1) shows the sub processes components of the company's information systems.

5. Project Tracking Register Information Systems (PTRIS)

The major objective of this submittal and correspondence information systems is to help decision makers, for management level (middle managers) and strategic level (senior managers) to monitor and plan for project information control strategies and company resource allocation and policies. The other objective of this system is to improve data accessibility within the company organization (departments and sections) and add value to project information received by central document control. This value represented in the following features:

- Assists in monitoring engineering team performance as an overall or for each discipline.
- Assess contractors' engineering progress (total predicted engineering document vs. submissions /approval to date).
- Produce instantaneous reports to be utilized at contractors/ consultant project management meetings.
- Refutes contractors' allegations of engineering review delays.

Figure (5.1) gives an overview idea about the relationships between the sub-processes and illustrates how the system functions.

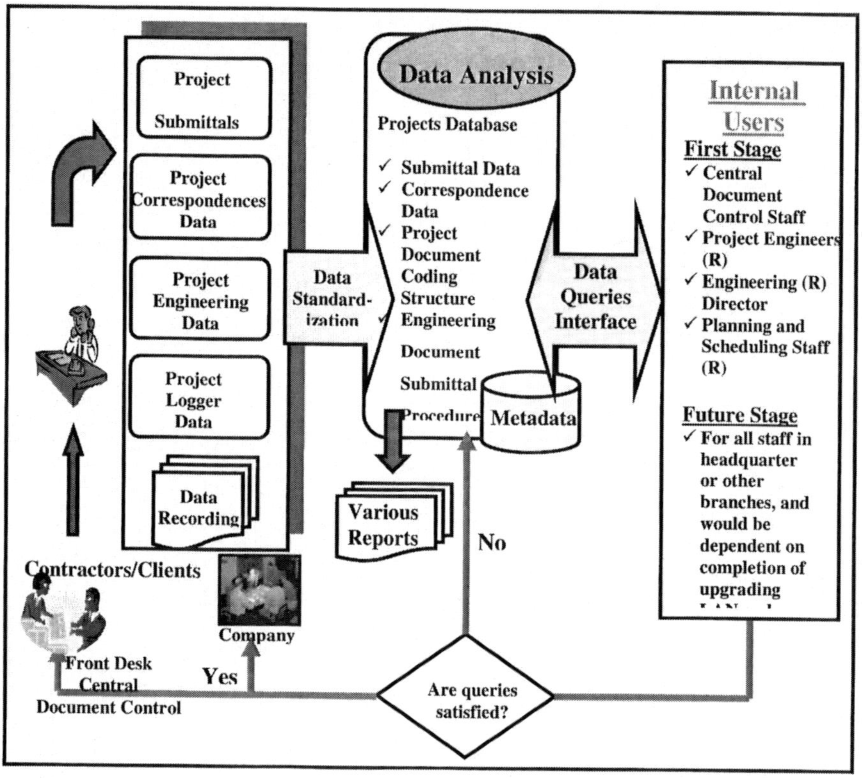

Figure (5.1) Project Document Tracking Register Information Systems Structure

6. Answering the User Information Needs

This item answers the users' information needs. Essential requirements of project information and the way forwards to be analyzed in the form of statistical reports is considered carefully and is used to produce the requested information (reports) by the user.

Produce a Summary Report of Engineering Document by Disciplines/Status, To-Date?

A select query can then be run by specifying the Discipline Code and Status.

This query analyses the project submittals data process and the results of this queries are a table as shown in Figure (6.1).

> The degree of accuracy and fast delivery of this report is dependent on the skills and technical abilities of the document controllers.

SUMMARY OF DOCUMENTS BY DISCIPLINES/STATUS FROM APRIL 2005 TO 05 MARCH 2007

DISC. CODE	DESCRIPTION	STATUS																			TOTAL
		AAC	AAP	ACA	AHO	ARR	CAP	CCA	CHO	CMT	CRR	HLD	IFI	INF	NCT	PAP	REJ	RR	SUP	WDN	
AA	Architectural				1	1	137	20		16	95		22	3	3		1				299
CC	Cost Control		3																		3
CR	Control		5	1		7	20	13			22					3		1			72
CS	Civil/Structural		40	49	76	68	22	29		30	67	2	1				56		2		442
CT	Contracts					1	1													3	5
CM	Construction Monitoring																				0
DC	Document Control		3	3						2											8
EL	Electrical		25	21	2	43	15	58	1	39	16		6	2		111	7			1	347
EN	Engineering General		8	5		4	6	10		1											34
GT	Geotechnical		29	8	1	14		1		1				2			1				57
HP	Building Services			8	1			1												1	11
HY	Hydraulics	1	2	13	2	15						1									34
IN	Instrumentation					19	9	12		1	17					4					62
MC	Corrosion Control		1	1			6	7			2						7				24
ME	Mechanical		5	64	12	1	15	7		1	21	10					7				143
MM	Materials Mgt/Procurement		12	11		4	3				1						4		1		36
PI	Piping			2	5	2	39			2	4			341	1		27				423
PR	Process			6		3	43	142		23	128					4					349
PS	Power Supplies					4	1	1		1	1		1			2	2				13
QA	Quality Assurance/Control	1	4	2		1	21	1			1						7			3	41
SA	Safety						1														1
SE	Scheduling/Planning		2			3				1							1				7
SY	Surveying			1			152	625			122			1						1	902
TC	Telecommunication		1				3	14		1	18										37
	Total per Status	**2**	**140**	**195**	**100**	**190**	**494**	**941**	**1**	**119**	**515**	**13**	**30**	**349**	**4**	**124**	**120**	**1**	**3**	**9**	**3350**

LEGEND

- AAC Approved with Comment
- AAP Approved for IFA
- ACA Conditionally Approved
- AHO Hold For IFA
- APC Approved Provided Comments
- CA Conditionally Approved
- CAC Conditionally Approved with Comments
- CAP Approved for IFC
- CCA Conditionally Approved-IFC
- CHO Hold for IFC
- CMT Comments
- CRR Revise & Resubmit for IFC
- FA For Approval
- GAP Approved as Built Drawing
- HLD Hold
- IFC Issue for Construction
- IFI Issue for Information
- INF Information Only
- IO Information Only
- N/R Not Required

Figure (6.1) Summary Table (Status vs. Disciplines)

What is the Monthly Progress of Engineering Review Process (Submittals vs. Responses)? , or What is the Engineering Trend?

With exporting (convert to excel) project submittals data to Ms Excel, Figure (6.2), in order to create an embedded chart on the worksheet, it is possible to see the engineering progress trend of all discipline or for particular discipline. This is important for senior management to assess the progress of engineering review process within the engineering department within a time periods. See Figure (6.3).

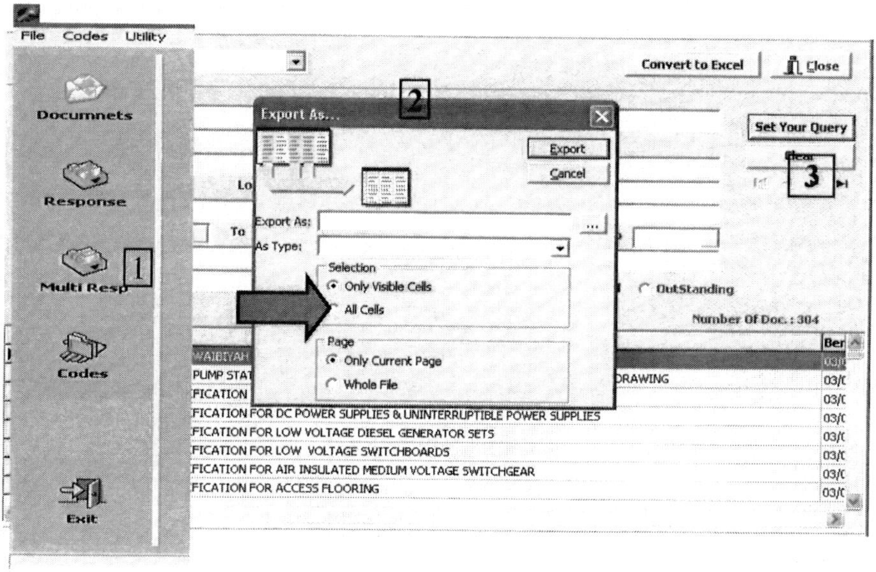

Figure (6.2) Exporting to Excel to produce extra reports

Month/Year	Apr-05	May-05	Jun-05	Jul-05	Aug-05	Sep-05	Oct-05	Nov-05	Dec-05	Jan-06	Feb-06
No. of Doc.	1	1	5	1	5	206	46	65	237	110	125
No of Respons	1	0	1	4	6	14	54	206	25	183	79

Month/Year	Mar-06	Apr-06	May-06	Jun-06	Jul-06	Aug-06	Sep-06	Oct-06	Nov-06	Dec-06	Jan-07	Feb-07
No. of Doc.	210	234	248	219	206	281	385	254	272	362	384	258
No of Respons	187	247	107	195	270	119	53	583	190	404	117	260

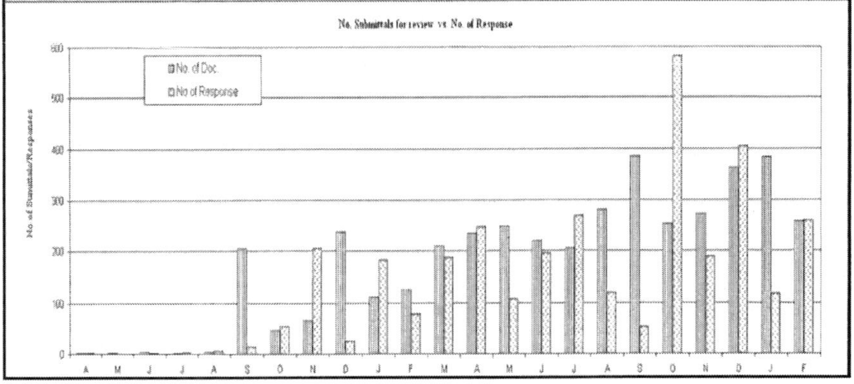

Figure (6.3) Monthly Progress Contractor's Submittals vs. Responses

Where are the Delayed or Overdue Documents for each Discipline?

This item answers the users queries pertaining to the overdue documents exceed the contractual review period, the result is a summary report sorted by disciplines.

A summary report can be viewed and printed via report options, overdue by discipline (summary report), then specifying the contract number and the constraint of review period. See Figure (6.4).

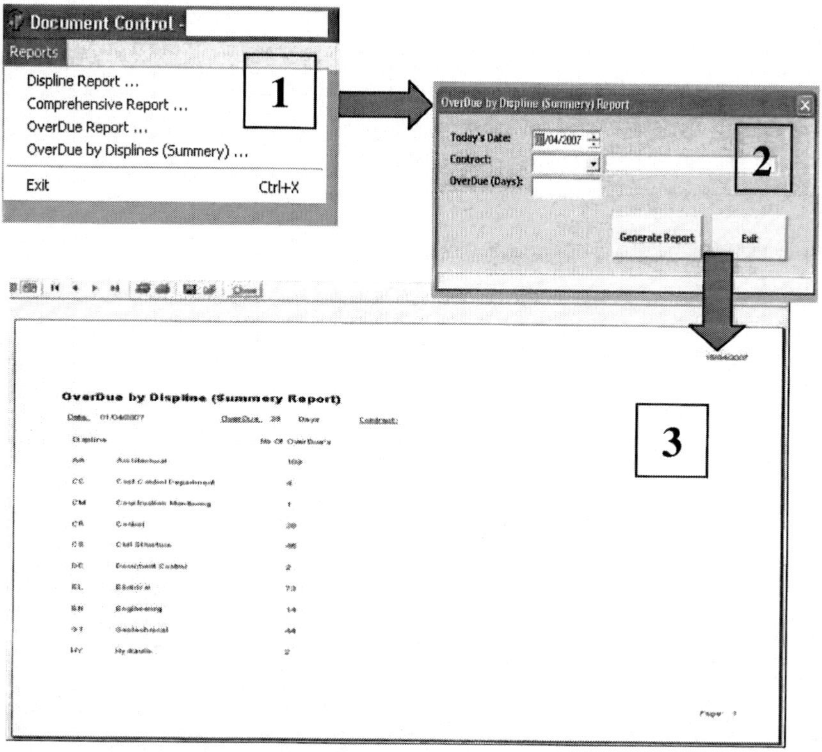

Figure (6.4) Overdue Engineering Documents, steps 1 to 3

Produce a Comprehensive Report for a Specific Contract/Project?

A production of comprehensive reports for any contract under the company's supervision can be produced by implementing a built in reporting facility within the systems, as a standard report. See Figure (6.5).

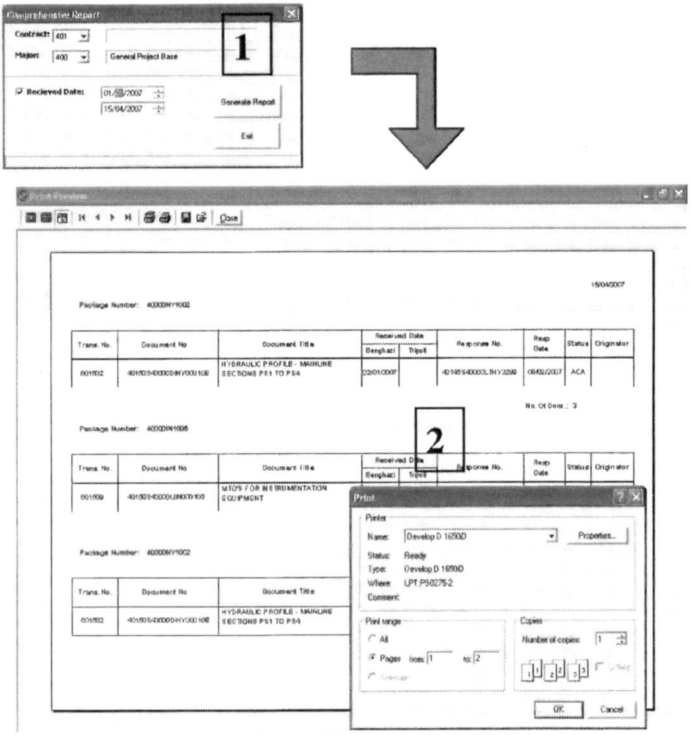

Figure (6.5) Production of a Comprehensive report, steps 1 and 2

Produce an Overall Graph to Reflect the Monthly Performance of Engineering Review?

With the exporting facility, the projects submittals data to Ms excel, then an overall chart could be updated and produced on a monthly basis, see Figure (6.6).

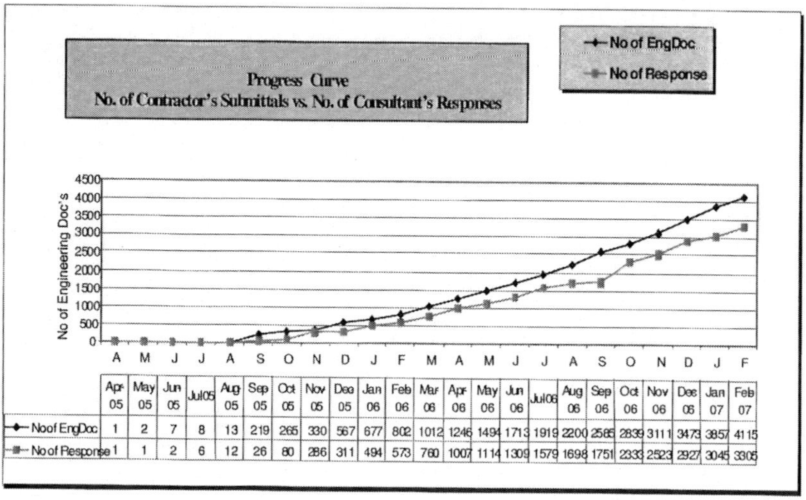

Figure (6.6) an overall Progress Chart

7. Old versus New PDTRIS

The designed and implemented PDTRIS will be utilized to register/control project management and engineering submittals and correspondence received from Contractors on a daily basis for approval/information. The below table summarizes the differences between the two tacking registers and advantages of the new registers.

Old PDTR vs. New PDTRIS

Old PDTR	New PDTRIS	Performance Objectives (Excelling)
2 Working under DOS environment that is no longer used/ supported.	Working under Window environment.	More **flexible**, More **dependable**, **Speed** the data entries, Enhance the **quality** of the outputs (reports

Old PDTR	New PDTRIS	Performance Objectives (Excelling)
		requested by the users).
3 Centralized systems, data controlled only by central document control staff, restricted access to others.	Decentralized, data could be controlled by document control and engineering However, the intention is to give access as read to all company's staff.	More **flexible,** Reduce the **cost**, as the engineering can produce his own internal reports, **Expedite** the flow of projects information within the company.
4 Impractical data entry screens, as it was designed for specific major project, and to satisfy contractual and organizational conditions.	More practical, as the data entries reflect the current requirements.	More **flexible** system, More **dependable** system.
5 Limited capability of production of projects reports.	Capable of producing the needed reports (various reports). The requested reports as standard or tailored, elaborated via excel or access.	Improves the **quality** of the outputs (reports needed by the users), Excels the **flexibility** of

Old PDTR	**New PDTRIS**	**Performance Objectives (Excelling)**
		reports presentation due to Ms excel features, **user friendly** application,
6 Designed to serve data entries of one project at a time.	Multi-projects system, data entries of projects data could be entered for any project at same time	More **flexible**, More **dependable**, **Speed** the data entries, Alleviate the **cost** impact; one document controller can manage more than one project.
7 There is no link between the front desk (central document control), and the engineering department.	The link is established via project engineering data sub-process. This grants the project engineers the abilities to monitor and control the flow of submittals within the engineering disciplines.	More **flexible**, More **dependable**, **Speed** the flow of information.

8. Conclusions

1. The major constraint faced by the Multi- National Co. being a lack of adequate information for projects to be exploited in decision making (i.e., response letters to Projects' owners and/or contractors).
2. Some information, were available, is not properly defined and its recording is not standard and does not satisfy the users requirements.
3. The company's Information technologies (IT) infrastructure is not-up to-date, local Area Network (LAN), and requires modernization.
4. New sources of data have been identified. The quality of data passing through the system has improved, because the recording of projects information data has also improved.
5. Components of information systems towards improvement have been identified and discussed as an important tool in project management information systems.
6. The new project document tracking register information systems (PDTRIS) has proven that it can satisfy most of the user needs and cope with the existing situation
7. There are actual and potential challenges that may be considered as a hampering factor for the full implementation of the database model. The linking of submittals data to get the required excel reports does not always result in complete matching, as this depends on the system's operator skills and technical ability. There is often the need for human intervention by way of re-matching the data interactively to put in relevant information forms.
8. This objective addressed the question of what provisions must be available for a successful implementation of the system designed. A number of conditions need to be considered such as skilled manpower with the right power level, evidence of what the system is capable of doing, financial resources, and specific responsibilities in maintenance of the information system components system are among the major conditions required.

9. Running of the project management information systems (PMIS) as a project may prove to be more efficient than incorporating it in the main stream of the organization. The bureaucracy may prove to be a stumbling block in its successful implementation.

References

[1] Company's Procedures 1997, Project Management System for Project Management, Administration and Finance, Personnel, Contracts, Engineering, Project Controls, Materials Management, Quality Assurance/Safety, and Project Engineering /Construction, unpublished.

[2] Awami, S. H. 2006. Detailed Specifications Requirements in Network Design for Benghazi Office, Internal Report, unpublished.

[3] Tiglao, N. Christopher, 2006. Module 10- GIS Design and Implementation. United Nations Statistical Institute for Asia and the Pacific (UNSIAP). Statistical Research and Training Center (SRTC). Available from: http://www.unsiap.or.jp/completed_prog/country/gis_phi_05/fpresentation/gism10.ppt [Accessed 1 January 2006].

[4] Camponovo, Giovanni, Ondrus J., and Pigneur Y. 2005. Environmental Context Significance in Strategic Decision Support Systems, International Workshop on Context Modeling and Decision Support. Paris. Available from: http://ftp.informatik.rwth-aachen.de/Publications/CEUR-WS/Vol-144/09_ondrus.pdf [Accessed 5 January 2006].

[5] Gnieber, O. K., 2005. Lecture Notes and Handouts in "Operations Management". Academy of Graduate Studies. Benghazi, Libya.

MANAGEMENT OF TECHNOLOGY SUPPORT CENTER FOR ENHANCING COMPETITIVENESS OF SMALL AND MEDIUM ENTERPRISES IN EGYPT

Yasser Tawfik, Competitiveness Research Group, Graduate School of Management of Technology, Nile University, Giza, Egypt*
Tarek Khalil, Competitiveness Research Group, Graduate School of Management of Technology, Nile University, Giza, Egypt**

ABSTRACT

The new global knowledge economy imposes numerous challenges for countries and enterprises in order to survive and stay competitive in the current dynamic and global marketplace. Rapid change of technology and aggressive competition in international market highlight the importance of creating new strategies to apply management of technology fundamentals including creating, acquiring and transferring of know-how with a view of integrating technology within the business model for enterprises. Egypt as a developing country considers Small and Medium Enterprises as an important component for the development of the Egyptian economy and for creation of jobs. In order for the Egyptian Small and Medium enterprises to achieve the country's economic development goals, they should pursue new and innovative technologies and follow progressive management strategies in order to stay competitive within the current global economy.

This paper proposes the establishment of a management of technology support center to provide a set of services to the small and medium enterprises in order to effect a sustainable economic development.

Keywords— Small and Medium Enterprise (SME), Management of Technology (MOT), Research and Development (R&D), Industry University Linkage (IUL), Venture Capital (VC), Cluster Model, Innovation.

INTRODUCTION

SMEs and Global Economy

SMEs have significant roles in development through: job creation that exceeds those of larger companies in most countries; outsourcing options for large firms; active contribution in services. SMEs represent high percentage of companies in the manufacturing sector in many countries, smaller firms are also increasingly present in technology-intensive industries such as information and communications technology (ICT) and biotechnology and they are involved in the innovation process by creating or re-engineering products or services and by introducing new organizational approaches.

Small and Medium Enterprises (SMEs) are one of the main engine for economic growth in developing countries and in emerging economies, This Drives governments to look for mechanisms to provide special support for SMEs initiation, growth and sustainability, in such a challenging dynamic globalized economy with strong competition in local and international markets. The knowledge-based economy leads to a new environment where effective management of technology, innovation and knowledge are the main assets for the creation of wealth, this create new ways of managing enterprises different from the ordinary ways of management.

Table 1. The role of SMEs in national economies (Percentages)

Country	Total Enterprises	Employment	Contribution to GDP
Australia	96	45.01	23.01
Canada	99.8	60.02	57.22
Denmark	98.8	77.8	56.7
France	99.9	69	61.83
Germany	99.7	65.7	34.9

Table 1. Continued

Ireland	99.2	85.61	40
Italy	99.7	49.02	40.5
Japan	99.5	73.82	57.03
Netherlands	99.8	57	50
Spain	99.5	63.7	64.35
United Kingdom	99.9	67.2	30.3
United States	99.7	53.7	48

Source: OECD [1], 1996

Table 1 shows that SMEs constitute more than 90 % of the total firms existing in many countries, constitute 45 % to 85 % of total employment, and contribute between 20% and 65% to the country GDP. SMEs are still facing many difficulties like obstacles for financing, inadequate knowledge in managing technological change, lack of high level management expertise and other challenges that differ according to the country. Egypt can depend on SMEs in creating new employment to face the tremendous population growth in the coming years.

SME AND GLOBLIZATION EFFECTS

Globalization has created a new business environment where large firms started to be more specialized by concentrating on core competence technology and capabilities while outsourcing non-core business components to other partners. Several approaches are used in this process including cross borders partners through business alliances, acquisitions, mergers and networking. SMEs can benefit from the new business environment through participating in strategic and cross border alliances and business networking. This is particularly evident in emerging technology fields and knowledge based fields like information and communication Technologies including Software, telecommunication, and computer networking related services.

However, SMEs specially the traditional ones in retail, services and manufacturing still can serve the local market, where creating a local market is also important.

Table 2. Software: cross-border marketing alliances involving SMEs, 1990-2000

	Software firm (SMEs)	Country	Employees	Alliance partner	Country	Employees	Year	Deal's highlight
1	Wind River Systems	Japan	208	Sun Microsystems	US	12 500	1991	Joint marketing of Wind River's computer operating system.
2	TA Consultants Ltd.	Hong Kong	80	IBM Corp	US	316 303	1991	IBM and TA established a joint venture to develop and market financial software in Southeast Asia.
3	Alias Research Inc	Canada	160	Silicon Graphics	US	3 624	1991	Silicon Graphics markets Alias Research's 3 D software products.
4	Unicom Systems Co	US	42	Integris (Bull)	France	10 000	1992	Integris markets Unicom's software.
5	Verbex Voice Systems	US	30	Mitsubishi Corp	Japan	13 959	1994	Verbex Voice Systems licensed Mitsubishi to manufacture and market Verbex Voice's software in Japan.
6	Syntax Systems Ltd.	Canada	170	Fujitsu Ltd.	Japan	145 000	1994	Syntax Systems Inc granted Fujitsu Ltd. a license to market Syntax's software in the US.
7	VideoLogic PLC	United Kingdom	140	NEC Corp	Japan	154 787	1994	NEC markets software jointly developed by the two companies.
8	FORE Systems Inc	US	10	Northern Telecom Ltd.	Canada	73 000	1995	The two companies co-market seamless end-to-end ATM solutions.
9	Elvis Ltd.	Russia	53	Sun Microsystems	US	12 500	1997	Sun Microsystems Inc markets Elvis Ltd's software worldwide.

Source: Sakai, 2004

As shown in Table 2, taking Software industry as an example we can see how SMEs join strategic alliances with multinationals like Sun Microsystems, IBM, Mitsubishi, Fujitsu and NEC. Also we see cross border alliances where SMEs from US serve large multinational companies in Japan and vice versa. Egypt software industry for example faces many challenges and need extra efforts to develop high quality SMEs who can take part in such quality driven global alliances.

Enhancing SMEs capabilities

To enhancing SMEs capabilities in order to compete globally and join high quality strategic business alliances, intensives efforts should be spent along various fronts. Selected areas where SMEs need help are discussed below:

Management Skills
SMEs which are going to search for strategic alliances and cross border business opportunities should improve the firm strategy in different fields especially toward enhancing management skills and hiring qualified human resources. SMEs Management should be responsive to the dynamics of the markets and customers needs and have the capabilities to search, find, negotiate and acquire international business opportunities. Enhancing management skills can be done through customized training, consultancy but also through practical assistance and practical training. New training fields are needed like management of technology, knowledge management, negotiation at international level, quality and supply chain management.

Access to strategic Information
Access to strategic information for global market is a big challenge for SMEs especially for developing countries such as Egypt. Business partnership opportunities and regulations in foreign market are example for strategic information needed by SMEs. Using ICT for information and electronic commerce is another need and a challenge for SMEs in the global market. Competitiveness is dependent on the ability of firms to acquire and apply new knowledge and technology in products and production processes in a timely manner.

Access to Business Networks
Business networks allows SMEs to get business information, initiate contacts with similar and larger firms in same fields, information sharing, and allows mutual learning between firms. Business networks are important at international and local level. Business cooperation can be initiated through the business

networks starting from informal business arrangement to business contracts. Business Networks can be through chamber of commerce or industry, trade associations, business matching services, Nongovernmental organizations Networks for enterprises and others.

There are more challenges and barriers for SMEs on the international level that can be seen in figure 1 shows results from an OECD survey indicating obstacles to internationalization facing SMEs and their ranking.

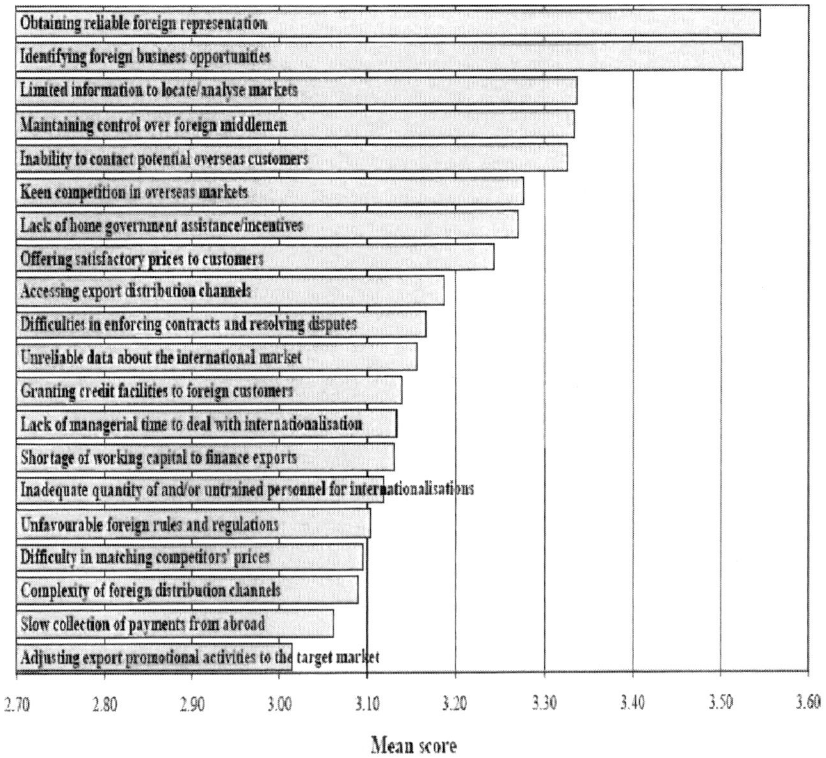

Figure 1. Obstacles to internationalization as perceived by SMEs, Source OECD, June 2007

Clusters

Clusters refer to geographic concentrations of interconnected companies and institutions in a particular field. They are informal linkages among firms and institutions, for example suppliers of specialized machinery and services, universities, standards-setting agencies, think tanks, vocational training providers, trade associations and others. A major characteristic of clusters is the geographical proximity of interdependent firms, which produces synergy effects. Close proximity allows firms in the cluster to transact more cheaply and easily, resolve their inter-firm problems more quickly and efficiently, and learn earlier and more directly about new innovations and business practices. Some successful clusters are strongly export-oriented, allowing small firms collectively to improve the quality of their products and achieve a significant share in foreign markets. (Sakai, 2004)

Innovative clusters can be defined also as networks of interdependent firms, knowledge-producing institutions (universities, research institutes, technology-providing firms), bridging institutions (*e.g.* technology extension services) and customers, linked vertically and/or horizontally in a value-adding production chain, and which co-operate in developing and using sector-specific public goods based on common physical and knowledge infrastructures. Clusters (innovative or otherwise) can contain small or large numbers of enterprises, as well as small and large firms in different ratios. They can be more or less knowledge-intensive, involving a comprehensive or more reduced set of knowledge-producing and bridging institutions, and have a narrow or broad sectored and technological focus since they occur in traditional as well as new industries. (OECD, 2004c)

Building innovative clusters in Egypt with focusing on enhancing SMEs in the local and international level is one of the main keys for economic development.

Table 3. Key Distinctions between Enterprise Clusters and Business Networks

Networks	Clusters
Allow access to specialized services at lower costs	Specialized Services to a region
Restricted membership	Open membership
Based on contractual agreements	Outcomes of market Dynamics.
Help firms engage in complex production	Generate demand for More firms with related capabilities.
Networks are based on co-operation	Require competition
Have common business goals	May have collective visions, including Public interest

Source OECD, June 2004c

There are differences between Networks and Clusters, Table 3. Shows the difference between Networks and Clusters where Networks are some kind formal networks with specialization and membership to gather people and firms who share the same interest. While clusters are some kind integrated service which bring different firms with different interest to form homogenous cooperation and it has open membership.

INNOVATION IN SME

Innovation is a key factor in firms' competiveness in the knowledge based economy. Innovative SMEs can be the main engine for creation of wealth in the global market today success depends on how effective innovation can be aligned within firm strategy. This requires building capabilities to develop, acquire, provide, implement and utilize new products, technology, process and services. Innovative SMEs face big challenges in finding finance for their innovation for different reasons such as: the uncertain returns of innovative activities; untested markets; the high cost of building prototypes and the overall development effort. There is also a need for new ways for technology and innovation assessments such as technology forecasting and technology valuation before investing.

New Ventures in general face challenges in financing innovation and require several rounds of financing. Each round can be financed from different sources. Table 4. Shows Amazon.com different financing cycle.

Table 4. Financing of Amazon.com (1994-99)

Time Line	Price/Share	Source of Funds
1994: July to Nov.	$0.001	Founder: Jeff Bezos starts with $10 000, borrows $44 000.
1995: Feb to July	$0.17	Family: Founder's father and mother invest $245 000.
1995: Aug to Dec.	$0.13-0.33	Business Angels: 2 angels invest $54 408.
1995/6: Dec to May	$0.33	Business Angels: 20 angels invest $937 000.
1996: May	$0.33	Family: Founder's siblings invest $20 000.
1996: June	$2.34	Venture capitalists: 2 venture capital funds invest $8 million.
1997: May	$18.00	IPO: 3 million shares issued raising $49.1 million.
1997/8: Dec to May	$52.11	Bond issue: $326 million bond issue.

Source OECD, June 2004a

SME AND INTERNATIONAL MARKET

Although there are some international opportunities for SMEs in the international markets, SMEs still show low representation in terms of countries exports. The later is still dominated by large firms and multinationals. A survey by OECD of European SMEs (2002) undertaken for 19 European countries found that 20% on average, of European SMEs were found to be involved in exports, the percentage varied by size class: micro enterprises (19%), small enterprises (32%), and medium-sized enterprises (46%). More

European SMEs in manufacturing were exporters (22%) than in services (14%). In the United States, the exporting role of SMEs (<500 employees) has been growing and two-thirds of exporting firms were firms having fewer than 20 employees in 1998. In contrast to European SME exporters, who exports in manufacturing sector, 70% of US exporting SMEs were outside the manufacturing sector.

In terms of their share in total exports, SMEs fare less well. In the United States, the SME share in total US exports averaged about 30% in recent years, 60% of which was accounted for by non-manufacturing companies. In Japan, the SME share (<300 employees) in total manufacturing exports was close to 50% throughout the nineties, similar to the Australian SME (<200 employees) experience, while in Korea, SME exports (<300 employees) varied between one third and two fifths of total exports during that time. The SME share in Korea's exports progressed from 36.9% in 2000 to 42% in 2002. (OECD, 2004b)

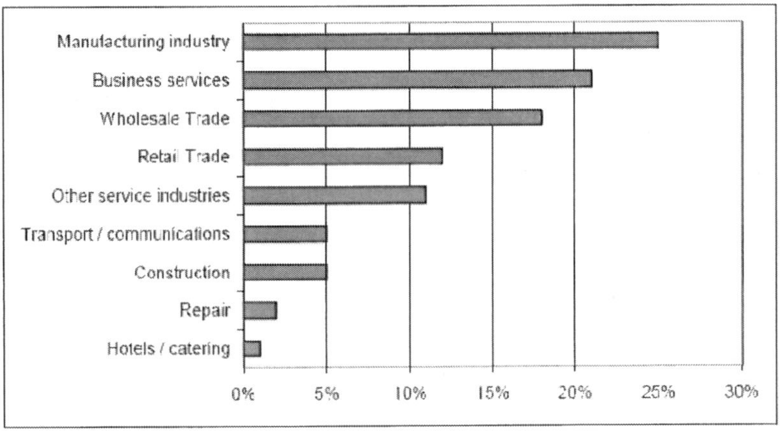

Figure 2. Percentage Share of European SMEs that Export, Europe-19, 2002, Source: OECD, June 2004b

Egypt lags in exporting high technology when compared to different countries and regions which shows the need for great efforts done by the government to enhance SMEs innovation and capabilities to contribute in the current globalized market. (Figure 3)

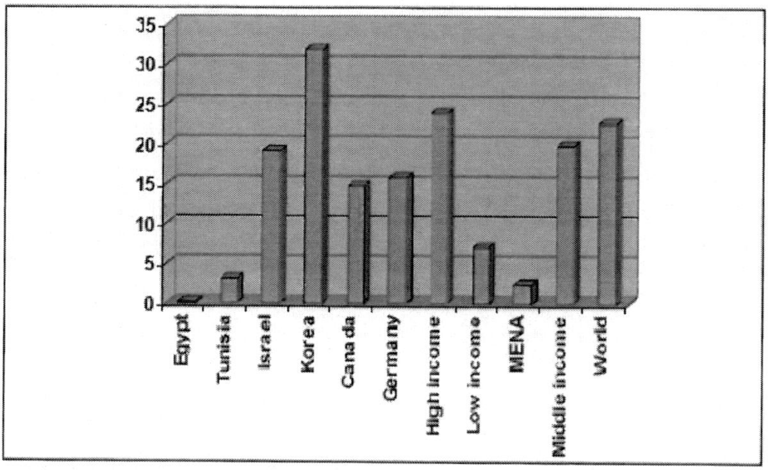

Figure 3. High Technology Exports as % of Manufactured Exports (1999), Source Ministry of Finance, 2004

SME IN EGYPT

As evident from the above review SMEs definition varies according to countries, organizations and objective. They are usually defined in terms of assets and/or number of employees. According to the European Commission recommendation 2003/361/EC: Micro enterprise are those that employ 1 to 9, Small enterprises employment 10 to 49 and Medium Enterprises employ 50 to 249 corresponding annual turnover is less than 2 million euro, less than 10 million euro and less than 50 million euro respectively.

In the United States SMEs are considered to be those employing less than 500 Employees.

The term SME or micro-enterprise is not commonly used in the United States, The US federal government has established the Small Business Administration ("SBA") (under the Small Business Act) and has a specialized bureau called The Office of Size Standards. This Office is tasked with defining and modifying the definitions of small businesses for purposes of federal government programs. The Small Business Act defines "...a small business concern as one that is independently owned and operated and which is not dominant in its field of operation. The Act also states that in determining what constitutes a small business, the definition will vary from industry to industry to reflect industry differences accurately. Most states (and even counties and cities) have small business development offices but do not use consistent definitions as they are uniformly in the business development field and not just confined to small businesses.

The SBA defines all enterprises according to their Standard Industrial Classification, (recently changed to the North American Industry Classification System, which standardizes classifications in Canada, the United States and Mexico,) category and sub-category in a document that runs to twenty-nine small print pages. The measures used are number of employees and total receipts and have a large range for both. For example, an aircraft manufacturing firm with fewer than fifteen hundred employees is considered small but durable goods wholesaler with one hundred one employees is considered large. Similarly, a gasoline station with a convenience store is a small business if it has gross receipts of less than $20 million whereas a gasoline station without a convenience store is small if it has less that $5 million in receipts. (Lerchs, 2001)

The definition In Egypt varies also according to organizations and objectives, for example the USAID funded Small and Micro Enterprise Development programs in Egypt define Micro Enterprises which is eligible for Loans range in size from LE 500 to LE 5,000 and have 1 to 5 Employees. The Egyptian ministry of industry definition for Micro Enterprises employment 1-9 Workers, Small Enterprises employment 10-50 and Medium Enterprises employment 50-100 with capital up to LE 500,000, LE 500,000-5 Million and LE 5- 10 Million Respectively.

The Credit Guarantee Company (CGC) is a joint venture between nine banks and an insurance company providing loan guarantees on behalf of SME clients, which are defined as being in any sector except trade and total assets, excluding land and buildings, of between LE 40,000 and LE 5 million.

When we compare the Egyptian Ministry of Industry definition for SMEs to The European Commission Definition we will find that they are quit closer to each others from the employment point of view except the medium enterprises upper limit is 249 in the EU definition while 100 only for the Egyptian definition where Medium enterprises in EU can reach the level of the Large firms in Egypt.

SMEs Activities in Egypt

SMEs account for 99% of all nonagricultural private enterprises in Egypt, account for the employment of more that 36% of the total labor force, 71% of the total SMEs are micro-enterprises in fact while the rest being divided between small (19%) and medium (10%).

Distribution of SMEs by activities in Egypt are 60.6% Wholesale and Retail Trade and Vehicle Maintenance, 16.8% Manufacturing, 5.1% Hotels and restaurants, 3.2% Real Estate, Renting & Business Services, 3.1% Health and social work, 2.4% Transportation, storage and communication and the rest for fishing, petroleum and mining, electricity, gas, water, construction, financial intermediation, education and community. The manufacturing sector is concentrated in few sectors. More than eighty percent of manufacturing enterprises are concentrated in five sectors. These in descending order are: Furniture, Garments, Food and beverages, Metal products (except machines), Wood & cork products. (Ministry of Finance, 2004)

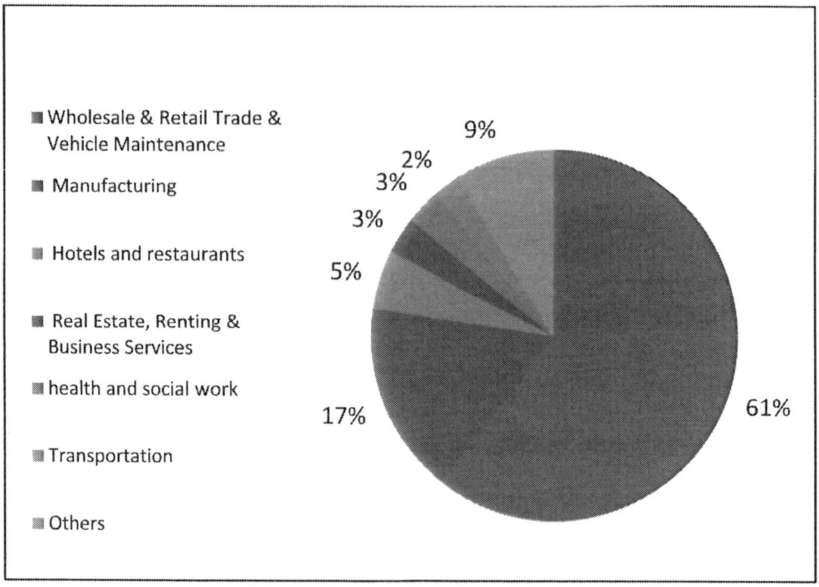

Figure 4. Distribution of SMEs in Egypt by Activity

According to CIDA Egypt Report about SMEs in Egypt, The national economic development strategy calls for the creation of 550,000 new jobs until 2017. At current growth rates of 3%, 513,000 additional jobs are required annually to cope with new entrants to the workforce. However between 1990/91 and 1995/96 the number of jobs created averaged only 330,000 jobs per annum. This does not absorb new job entrants or reduce unemployment and underemployment, which would require net job creation in the economy of 600,000 jobs per annum. The two largest employment sectors agriculture (31%) and public utilities (18%)-are unlikely to contribute to higher rate of job creation as both suffer from excess employment and low productivity. If the job creation target is to be achieved, much will depend on the absorptive capacity of SMEs. (Ministry of Foreign Trade, March 2002).

CHALLENGES FOR SME COMPETITIVNESS IN EGYPT

SMEs in Egypt face many challenges beside the international ones mentioned earlier for all SMEs. Some examples are mentioned below:

Technology and Research and Development (R&D)

SMEs in Egypt should upgrade their machines and equipment and take the advantage of existing technologies, introduce new technologies and new production and management process. Innovation and R&D lack sufficient funding, Management capabilities in planning and the know-how of integrating innovation and R&D strategies in the business models. Industry-university-research center linkage in the performance of joint researches and cooperation is also very low.

Access to Finance

SMEs and especially innovative SMEs have difficulty accessing finance at different phases of the technology development and deployment lifecycle. A major Problem is the small amounts of Venture Capital funds especially technological VC; lack of sufficient angel funding and funding for prototyping or prove of concept. The Egyptian Social fund for Development provides some supports but usually inadequate and mostly in low-tech based ventures.

Access Local and international Markets

Accessing local and international markets for the Egyptian SMEs are limited by the availability of adequate and accurate information about markets and inaccessibility of Exports channels. Business to business linkage to local and international firms is very weak.

Need for Uniting Efforts

There are many governmental, NGOs, Private initiatives and programs for supporting SMEs but no one source for collecting these programs information, little coordinating efforts between different initiatives which lead to repetition of programs. There is also a recent interest in promoting entrepreneurship but still in early phases of development.

RECOMMENDATION FOR ENHANCING SMES INNOVATIVE CAPACITY

Nile University Competitiveness, Entrepreneurship and Innovation Center (NU CEIC)

Based on analyzing the SMEs information mentioned in this paper, a cluster model, new to Egypt is recommended; it is based o providing services through Nile University Center for enhancing SMEs competitiveness and innovative capacity. New challenges and opportunities are present in the knowledge age for the Egyptian private, public, financial and academic sector and for organizations involved in the field of technology and industry. SMEs establishment, capitalizing on innovation and technology opportunities can be created at the intersections of these sectors networking through a collaboration entity. One of the most promising areas of a cluster in Egypt is technology-based cluster particularly concentrating on the Communication and Information Technology (CIT). The Creation of the Smart Village in Giza, Egypt is a golden opportunity for the creation of such a cluster model. Smart Village is Egypt's premier model of Public-Private-Partnership investment with the strategic aim to lead and foster the development of profitable chains of Technology and Business Parks. Currently, 12,000 professionals run the operations of more than 100 Companies and expected to reach 80,000 by the end of 2014. (Smart Village, 2008)

Nile University (NU) can play a significant role as the institute of collaboration and as a catalyst to build a successful innovation cluster model (Figure 4). A Nile University Competitiveness, Entrepreneurship and Innovation Center (NU CEIC) can be established with its main purpose to provide: Research, services, consultancy and networking for SMEs and the other components of the innovation cluster model. The Center can provide services needed for enhancing SME Competitiveness.

The NU CEIC Main Services are:

(i) Consultancy in Technology, Business and marketing.
(ii) Professional Training.
(iii) Networking.
(iv) R&D support to increase value added.

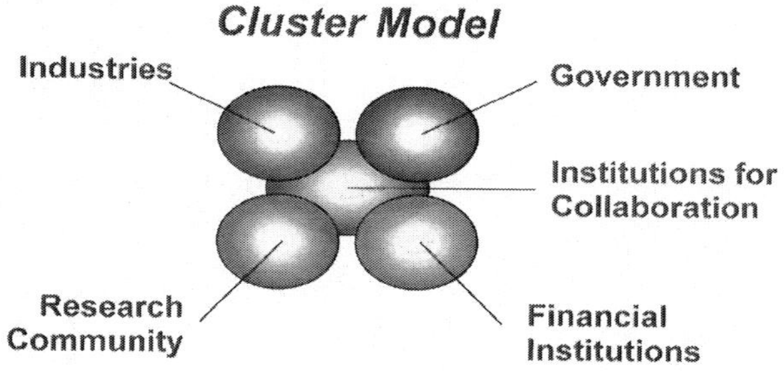

Figure 5. Cluster Model for Innovation, Source Gatti, 2007

Services

(i) Industry (SME)

SMEs in field of technology and industry need new ways of management besides the traditionally management used by ordinary companies. This needs wide area of knowledge, experience, programs adaptation and customization per company. The NU CEIC will provide Research, professional and customized training and consultancy about how to do:

(a) Planning and implementing innovation strategies in organizations.
(b) Technology Forecasting.
(c) Product Design and Development.
(d) Technology marketing.
(e) Competitor analysis.
(f) Innovation management.
(g) Production Technologies and management.
(h) Core competency determination.
(i) Linking Technology Strategy and Business Strategy.

(j) Technology planning and road mapping.
(k) Technology auditing: Technology environment, categorization, markets and competitors, Innovation process, value added functions, acquisition, exploitation and protection.
(l) Technology transfer.
(m) Customer Focus Strategy.
(n) R & D strategies.
(o) Technology and its enablers: management of technological resources, interface with external environment, management of R & D and engineering projects, management of Human resources under conditions of technological and social changes, organizational structure in times of high competitive pressure.
(p) Developing a competitive strategy: The strategy should be a broad formula including the Basis on which the firm is going to compete, the goals of the business and the policies needed to reach these goals and consultancy for competitive strategy including Target Market, Product Line, Marketing, Sales Distribution, Manufacturing, Labor, Purchasing, R&D and Finance &Control. (Khalil, 2000)

Research community
NU CEIC can help research community in Egypt to be more linked to industry and commercialize their research efforts for the benefit of the industry and the communities through providing know how about building:

(i) Research and development programs and strategies
(ii) Successful Commercialization Program
(iii) Full Business Incubation systems
(iv) Industry University research collaborative centers

Government
NU CEIC researches and consultancy can work close with decision makers In the government to develop research and development

strategies, competitiveness strategies, innovation clusters models for Egypt in general.

Financial institution
The Ministry of finance analysis for Venture Capital Funding in Egypt reported that there are currently a small number of venture capital firms in Egypt. However, seeking quick turnaround of their funds, they look for investments that could be floated in the stock exchange within a two-to-three year period. These terms while possibly suitable for growing enterprises do not meet the needs of new start-ups like they do in developed, and several developing countries. In addition, innovative companies which need a long growth period before they become marketable are hence an unattractive target for the venture capital industry in Egypt. Finally Egyptian SME owners, like their counterparts in many countries, are reluctant to give up equity in returns for investment. (Ministry of Finance, 2004)

NUCC will work with Current Venture Capital funds to develop their vision toward technology based SMEs and how to do technology valuation and intellectual property rights issue in order to start invest in the technology based SMEs. One of the main activities that NUCC can do is networking of the research community, SME in Egypt with international venture Capitals which will help in:

(i) Provide new channels for funding
(ii) Provide access to international market through venture capitals in different countries, as it is difficult to enter different countries market without a knowledge partner.

Collaboration
The innovation model shown in Figure 4 shows Institutes of collaboration which are the main drive for making the innovation cluster model successful in Egypt to unite efforts between all other entities. NUCC will be the catalyst for achieving collaboration between all the above entities to achieve successful innovation cluster model for SMEs in order to compete globally.

CONCLUSION

Small and Medium Enterprises (SMEs) are the main engine for job creations and economic growth for countries. The Knowledge based economy provides opportunities and threats for SMEs worldwide. Opportunities for SMEs arise from the new business environment characterized by growth of outsourcing and supply chain management. Many firms are concentrating on core competence technology and capabilities while outsourcing the other business components to partners including cross borders partners through business alliances, acquisitions, mergers and networking SMEs in different countries. Innovative SMEs can also benefit from the added value solutions to compete in the international market. Threats arise from the rapid change in knowledge and technology and aggressive competition from large and multinational companies in the local and international market. SMEs in Egypt can play significant roles in developing the country economy and increasing job creation. There is a serious need for initiating special support programs for innovative SMEs by government, academic institutions, nongovernmental organizations and professional associations. Such Support is needed to improve SMEs potential for success to be able to compete and sustain growth in the knowledge based economy. Clustering and networking models can enhance SMEs competiveness in Egypt. A recommended cluster model for innovative SMEs in Egypt is proposed with collaboration in education and services delivered through Nile University Competitiveness, Entrepreneurship and Innovation Center (NU CEIC). SMEs coordination in Egypt faces some challenges. The most important are the unavailability, credible updated data and information, and the challenge of coordinating many institutions involved with SME development but had historical difficulty in synergizing efforts.

REFERENCE

Gatti, Gabriele, 2007, "Innovation clusters in Italy. Managing innovation and internationalization: The experience of AREA Science Park", Sino-European Forum on Innovation Clusters, Maisons-Alfort.

Khalil, Tarek, 2000, "Management of Technology", McGraw-Hill.

Lerchs, George, September 2001, "The study of an operational definition for Micro, Small and Medium Sized Enterprises in Egypt", The Canadian International Development Research Center (IDRC) Contract SMEPOL/100168.

Ministry of Finance, 2004, "Enhancing Competitiveness for SMEs in Egypt", General Framework & Action Plan.

Ministry of Foreign Trade, March 2002, Streamlining the Regulatory Procedures for Small and Medium Enterprises "The One Stop Shop Model", the Small, Medium and Micro Enterprise Policy Development Project (SMEPOL) Publications.

OECD, 1996, the Organization for Economic Co-operation and Development (OECD), www.oecd.org, "SMEs: Employment, Innovation and Growth", the Washington workshop.

OECD, June 2004a, "Financing Innovative SMEs in a global Economy", 2nd OECD conference for SMES promoting entrepreneurship and innovative SMEs in a global economy: towards a more responsible and inclusive globalization.

OECD, June 2004b, "Facilitating SMEs access to International markets", 2nd OECD conference for SMEs promoting entrepreneurship and innovative SMEs in a global economy: towards a more responsible and inclusive globalization.

OECD, June 2004c, "Networks, partnerships, clusters and intellectual property rights: Opportunities and challenges for innovative SMEs in a global Economy". 2nd OECD conference for SMES promoting entrepreneurship and innovative SMEs in a global economy: towards a more responsible and inclusive globalization.

OECD, June 2007, "Enhancing the Role of SMEs in Global Value Chains", OECD global Conference Tokyo.

Sakai, Kentaro, 2004, Directorate for Science, Industry issues global industrial restructuring: "Implications for Small firms Technology and Industry", the Working Paper series of the OECD Directorate for Science, Technology and Industry

Smart Village, 2008, www.smart-villages.com.

PART 6

" Community "

- *authorities , hubs and brokers in communities of practices*
- *Does fertile environment have positive effect on university technology transfer activities? : The case study on two Texas state universities*

AUTHORITIES, HUBS, AND BROKERS IN COMMUNITIES OF PRACTICES

MARIANNE HÖRLESBERGER*

Austrian Research Centers GmbH - ARC; systems research, Donau-City-Straße 1, A-1220 Wien, Austria, marianne.hoerlesberger@arcs.ac.at

PETRA WAGNER-LUPTACIK

Austrian Research Centers GmbH - ARC, systems research, Donou-City-Straße , A-1220 Wien, Austria petra.wagner-luptacik@arcs.ac.a

*) *corresponding author*

Abstract: Communities of Practice (CoP) have gained importance as a key knowledge management instrument for knowledge sharing in organisations. A CoP is defined as a group of people bound by informal relationships and who share common practices. In this contribution we investigate CoP within a research organisation as knowledge networks where a researcher is considered a node in a knowledge network graph and his/her communication links to others in the CoP as edges. We focus on identifying "important" nodes for knowledge flows such as authorities, hubs and brokers which can be identified by looking at directed graphs. Those CoP members to whom large numbers of people go for advice are authoritative sources. A node with a high in-degree is a good candidate for becoming an authority. A researcher expert with a corresponding high out-degree has a high potential of becoming a hub. Hubs are understood as nodes with high distributive capacity. Knowledge flows are represented as a directed network graph. We calculate the network graphs in this contribution with the software Pajek. For identifying authorities and hubs we apply Kleinberg's Algorithm. For identifying brokers/bridges we apply the measure of betweenness centrality, which measures the shortest path of an expert to all others in the network (developed by White&Borgatti for directed graphs). Experts with a high betweenness centrality have the potential of bridging parts in the network. We link the approach of White&Borgatti with the idea of Kleinberg for detecting researchers with authoritative, hub or broker roles in the CoP knowledge network. Interpreting only the results of the quantitative data without knowing the background of the data could lead to skewed conclusions. Thus the quantitative analysis was complemented with expert interviews in the research organisation. Our results were confirmed and they are helpful but only together with the background information of each node. The introduced methodological approach can be considered as being adequate.

Keywords: Communities of Practice, network analysis, identifying important nodes, hubs and authorities, brokers.

Introduction

Organisational knowledge sharing enhances the potential of an organisation for innovations. Research organisations especially depend on knowledge sharing of their employees. Communities of Practice (CoP) have gained importance as a key knowledge management instruments for knowledge sharing. A CoP is defined as a group of people bound by informal relationships and who share common practices in their field of work. Members of CoP also help each other in solving problems. The importance of CoP has been worked out in a number of contributions by Lave and Wenger (1990), Brown and Duguid (1997), and Boland and Tenkasi (1995). In this contribution we investigate CoP in a research organisation by conceptualizing them as knowledge networks. More specifically, a researcher is considered as a node in a knowledge network and his communication links to other members of the CoP as directed edges. We focus on identifying "important" knowledge nodes (experts) and their particular roles in the network.

Firstly we give some background about CoP and a clue for the applied analysis methods. Secondly we introduce the investigated data and applied algorithm and measures. Thirdly we present the results in chapter Outcome. Conclusions complete our contribution.

Background

McDermott (1999) pointed out that new knowledge is created at the boundaries of established knowledge. Particularly research organisations are composed of experts who share a strong interest in a specific topic and work at the frontiers of science to create new knowledge. Communities of practice have increasingly been recognized as key to effective knowledge production as person-embodied knowledge is increasingly perceived as competitive strategic asset in organizations (Wenger and Snyder 2000). Communities of practice are "groups of people who share a concern, a set of problems, or a passion about a topic, and who deepen their knowledge and expertise in this area by interacting on an ongoing basis" (Wenger et al. 2002, p.4). From the first studies on situated learning (Lave and Wenger 1991) and its management interpretations (Brown and Duguid 1991), communities of practice approaches see knowledge generation as essentially a social process through which individual and organizational learning takes place. Moreover, identities are formed in part through becoming an accepted member of a group. Communication as social process connects the community and its members. The definition of the term "practice" according to Wenger focuses on these communal resources that form the basis of communication within a community: "It (practice) denotes a set of socially defined ways of doing things in a specific domain: a set of common approaches and shared standards that create a basis for action, communication, problem solving, performance and accountability." (Wenger et al 2002, p.38). Continuous participation generates identity and trust in the community and enables community members to generate tacit knowledge quickly among each other.

Our study investigates a Community of Practice in a contract research organization, where the management initiated CoPs to leverage methodological know-how residing in different organizational subunits. CoPs aim at improving cross-functional networking within the research organization – both for scientific knowledge exchange and co-development of new project ideas and cooperation. The methodology applied is described in the following chapter.

Methodology

Data on work-related communication and knowledge links among the members of the CoP was collected with a questionnaire. The experts (members of the CoP) were given several questions with the list of the members of the CoP in the sense Cross and Parker suggest [Cross, Parker (2004)]. We analyse the two questions:

(i) The communication network (Q1): How often do you talk with the following people regarding your work? (With the options to mark "never", "sometimes", "often", and "consistently").

(ii) The knowledge network (Q2): "I understand this person's skills and knowledge independently of their relevance for my work". (With the options: "none", "moderate", "good", "very good").

The answers and results to the first question can be presented here in detail. The frame of this article does not allow presenting the results of the second question, but were presented during the IAMOT conference. Nevertheless mentioning it here means that question number 2 is relevant for tracing the information flow.

Knowledge flows are represented as a directed network graph, calculated with the software Pajek (http://vlado.fmf.uni-lj.si/pub/networks/pajek/). Each expert is represented as a node in the network visualisation. The link between two nodes is called a "directed edge", or an "arc". The width of an arc corresponds to the frequency of interaction:

"Never", or "non" gets the value "0",

"Sometimes", or "moderate" gets the value "1",
"Often" or "good" gets the value "2", and
"Consistently" or "very good" gets the value "3".

The members of the CoP at hand are affiliated to different departments. Each node is visualised in a specific geometrical form so that we can see the cooperation patterns among departments.

We define the in-degree of a node as the sum of all values of all incoming arcs. That means if the expert (node) X is attributed the value "1" from expert (node) A, value "0" from expert (node) B, and "3" from expert (node) C, then the absolute value of incoming arcs of expert X is the sum of all these values, namely "4". The out-degree is defined analogously.

For identifying authorities and hubs we apply Kleinberg's algorithm (1998), originally developed for detecting characteristic nodes (URLs) in the World Wide Web which is also a directed graph. Kleinberg mentions in his article that his algorithm can be applied for social networks, too.

Kleinberg's algorithm

Although the Kleinberg's Algorithm is explained in great detail in (Kleinberg, 1998), the most important formulas are presented here once more for a better understanding of the application.

(p,q) Be a directed edge from the node p to node q. Be E the whole considered set of the directed edges.

$a(p)$ Be the authority score of p

$h(p)$ Be the hub score of p .

Then operation I for each node p is given via:

$$a(p_r) = \sum h(q_s)$$

For (p_r,q_s) of E, respectively for $a(q_s)$. r goes through all indices of p and s through all indices of q.

Operation O for each node p is given via:

$$h(p_r) = \sum a(q_s)$$

for (p_r, q_s) of E, respectively for $h(q_s)$. r goes through all indices of p and S through all indices of q.

Let A be the adjacency matrix. An entry for a pair (p_r, q_s) is 1(If there is a link the originally Kleinberg Algorithm allocate it to the value 1. We consider here the values $1, 2$ or 3 for possible arcs, a connection from one node to another) if p has a link to q, else the entry is 0. Let h_i be the vector of hub scores after i iterations and a_i be the vector of authority scores after i iterations. The operation after i iterations is:

Operation I: $\quad a_i = A^t h_{i-1}$

Operation O : $\quad h_i = A a_i$.

All authority and hub scores are normalized after each operation via:

$$a^*(p) = a(p) / \sqrt{\sum_{q \in V} [a(q)]^2}$$

$$h^*(p) = h(p) / \sqrt{\sum_{q \in V} [h(q)]^2},$$

Where V is the set of authority weights respectively hub weights of the considered iteration before the normalization.

Kleinberg starts with the authority source in his algorithm and iterates the values of authorities and hubs interactively between the authority and hub sources. The iteration reinforces the authority and hub candidates. The values converge to approximation values in the runs of the iterations. In our case the process is stopped when the characteristic development of the values is recognisable. This can be done because it is proven that the algorithm is converging.

We apply Kleinberg's algorithm in the following way. We start the iteration with the absolute values of the incoming links (i.e. the authority candidates) as explained above and normalize them according to Kleinberg's algorithm (see the formula $a^*(p)$). These results represent the input score for the first operation for calculating the hub nodes.
Authority and hub values range between 0 and 1. High values indicate the best authorities, or hubs respectively.

Betweenness centrality
For identifying brokers/bridges we apply the concept of betweenness centrality. Betweennness centrality measures the shortest path of a node (in our case an expert) to all others in the network. Freeman (1977, 1979, 1980), Wassermann&Faust (1994), and Cross (2004) defined betweenness centrality in undirected graphs, and White&Borgatti (1994) for directed graphs.

Betweenness centrality is a measure of centrality that considers the position of nodes in between the geodesic (i.e. shortest path) that link any other node of the network. Considering directed graphs we have to deal with non-symmetric relations. If node i (an expert in our case) links to j, we say j is adjacent from i. A set of arcs linking i to j to k determine a path from p_i to p_k. There can be more than one path for any ordered pair of nodes.

For directed graphs we have to distinguish between a reciprocal link and an edge which goes only in one direction. If n_O is the number of nodes with outgoing arcs and n_I is the number of nodes with incoming arcs, then in a star graph the product $(n_I - 1)(n_O - 1)$ is the number of paths through the center. Let n_S be the number of nodes with reciprocal arcs. Then the betweenness centrality of the most centralized star of a directed graph is given by

$$C_B(p_i) = (n_I - 1)(n_O - 1) - (n_S - 1),$$

Whereas $0 < n_S < n$, and n the total number of nodes.

The relative betweenness centrality of a node k is its betweenness centrality divided by the maximum centrality, holding constant the number of nodes in the graph with reciprocal arcs n_S or only reciprocal arcs $(n_I - n_S), (n_O - n_S)$ given by the formula:

$$C'_B(p_k) = \frac{C_B(p_k)}{\max C_B(p_k)},$$

This is the same as

$$C'_B(p_k) = \frac{C_B(p_k)}{(n_I - 1)(n_O - 1) - (n_S - 1)}.$$

The value lies between 0 and 1. Nodes (experts) with a high value of betweenness centrality have potential of bridging parts in the network graph.

Outcome

We calculated network graphs of the investigated CoP. The various graphs (Figure 1 to Figure 7) represent the same CoP, but analyzed from different viewpoints.

The geometrical form and colour of a node provides information about the department to which the expert (represented in the considered node) belongs. The size of a node represents the value of the in-degree, what means the higher the in-degree, the larger the node.

The communication network (Q1):

As described in the methodology section, the members of the CoP provided information on communication: "**How often do you talk with the following people regarding your work?".** They had to mark the following values for frequency: "never" (=0) "sometimes" (=1), "often" (=2), and "consistently" (=3). These values determine the width of the arcs (connections) from one node to another in the network graph (Figure 1).

Figure 1 shows that there are two departments strongly connected in the centre of the graph. These departments are linked by several "bridges" to two other departments in the "west" and to one in the "east" of the graph. Two nodes are isolated in the "south" of the network graph, but with a reciprocal connection. Nodes with high in-degree are rather in the centre of the graph. The department located in the south of the centre shows the largest nodes, which means nodes with the highest in-degrees; and this department is also that one with the highest number of CoP members. The strongest links can be detected within the two departments in the middle of the graph and also between them. The department in the east with the nodes 15, 41, 48 is linked strongly internally, similarly the department in the North West with the nodes 84, 88, 42, and 9.

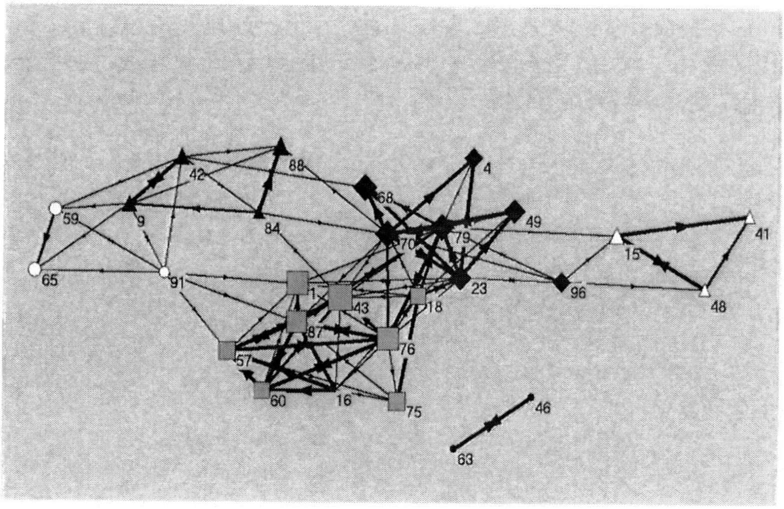

Figure 1. Communication network graph of the CoP (Q1), calculated with Pajek.

In the following, we analyse authorities and hubs. For understanding the analysis, we have to describe the data set in more detail. There were 28 members in the CoP. However, there were only 27 members listed in the questionnaire (because of a mistake). Therefore we have only 27 nodes as starting score for the authority candidates (because an authority is based on incoming links). 19 members completed the questionnaire. These 19 members build the base score for calculating the hubs (because a hub is based on outgoing links). Betweenness centrality, however, can be calculated for all 28.

Now we take a closer look at the authorities in our information network graph (Q1). An authority is a person with the power to influence people because they respect her/his knowledge or official position, or is a person with special knowledge. Concerning network analysis and Kleinberg's algorithm, the authority candidates correspond with the incoming links. The base score for calculating the authorities is the sum of the intensity, the so called in-degree, by which a person is asked or contacted concerning work ("never" = 0, "sometimes" = 1, "often" = 2, and "consistently" =3). Figure 2 presents the visualisation of this information network. The five nodes with the highest authority values are marked. They are located in the centre part of the network graph. Four of them belong to the same department in the south of the graph, which is also the department with the highest number of members in the CoP.

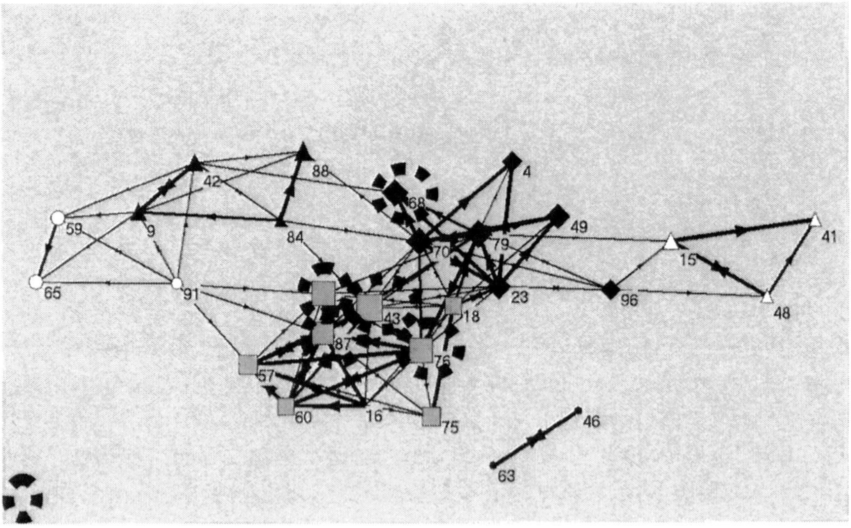

Figure 2. Network graph of the CoP to Q1, nodes with the highest authority values are marked.

Nodes with the highest authority values are marked like that.

Table 1 shows that node 43 (i.e. expert with number 43) ranks first (0.4368 after the third iteration). The second rank is on node 76 (0.3824), followed by node 1 (0.3457). Node 87 ranks fourth (0.2925), followed by node 68 (0.2362). Expert 68 is the only one of the top five affiliated to the department located in mid north of the graph (see Figure 2), because the other four belong to the department represented in the mid south part of the graph.

Table 1. The authority ranks for the communication network (Q1) with the three steps of iteration to each node and their in-degree ranking.

Authorities	1. Iteration	2. Iteration	3. Iteration	Rank	In-degree	Node
a (43)	0.4110	0.4349	**0.4368**	1	17	76
a (76)	0.3653	0.3735	**0.3824**	2	16	87
a (1)	0.3197	0.3548	**0.3457**	3	14	43
a (87)	0.2740	0.2983	**0.2925**	4	13	57
a (68)	0.1598	0.2289	**0.2362**	5	12	60
a (49)	0.2055	0.2129	**0.2251**	6	11	79
a (79)	0.1827	0.2150	**0.2198**	7	10	68
a (18)	0.0913	0.2012	**0.2133**	8	9	49
a (75)	0.1370	0.2124	**0.2095**	9	9	1
a (70)	0.2740	0.2070	**0.2037**	10	9	70
a (23)	0.1598	0.1809	**0.1855**	11	8	23
a (57)	0.2055	0.1910	**0.1769**	12	7	75
a (4)	0.1142	0.1611	**0.1730**	13	7	4
a (96)	0.0685	0.1611	**0.1730**	14	6	88
a (60)	0.2283	0.1606	**0.1573**	15	6	9
a (88)	0.0913	0.1089	**0.0926**	16	5	42
a (15)	0.1370	0.0854	**0.0799**	17	5	18
a (9)	0.1142	0.0747	**0.0533**	18	5	41
a (59)	0.2055	0.0688	**0.0419**	19	5	15
a (65)	0.1827	0.0582	**0.0325**	20	4	48
a (42)	0.1598	0.0432	**0.0260**	21	4	65
a (84)	0.1142	0.0347	**0.0184**	22	3	46
a (48)	0.0685	0.0235	**0.0161**	23	3	63
a (91)	0.0457	0.0240	**0.0090**	24	3	96
a (41)	0.1142	0.0101	**0.0028**	25	3	59
a (46)	0.0685	0.0016	**0.0000**	26	2	84
a (63)	0.0228	0.0005	**0.0000**	27	2	91

Table 1 shows that the values of the three iterations per node increase for the first eleven ranks (from a (43) to a (23)). We can say that the authority candidates are strengthened. From rank 12 onwards the values of the three iterations per node decrease.

Taking a closer look at the table of the in-degree (Table 1, right part) we realise that this ranking differs from the authorities ranking. The first five ranks are taken from the nodes 76, 87, 43, 57, and 60 in order. Here node 43 ranks third and not first. The nodes 57, rank four here, and 60, rank five here, are on the places 11 and 15 in the authorities ranking. Why can this happen?

This is caused by Kleinberg's algorithm. It always takes the previous hub scores for the next iteration step of an authority. Let us consider node 60. The in-degree of node 60 is 12, the sum of the values of the incoming links, and ranked of five. Its authority rank is 15. Kleinberg's algorithm calculates the authority value of node 60 via

$$a(60) = h(16) + h(57) + h(76) + h(87),$$

Whereas the values of the hubs come from the previous hub iteration.

We see here the hub values of the nodes adjacent to node 60 are responsible. And these values are not high enough for node 60 to achieve a higher rank.

Another reason that the ranking of the authorities and the in-degree differ is that only 19 of 28 experts responded to the questionnaire. As an authority value is calculated by the hub values of its preceding iteration, the chance for achieving a higher value depends on the incoming links of the active nodes, in this case the hub candidates, i.e. the ones who answered the questionnaire.

The in-degree provides a first orientation of important authority nodes. The authority value takes also into account the importance of the nodes which link to the node in question. Now we consider the hub nodes. The term "hub" means being the central and most important part of a particular place or activity, e.g. the commercial hub of the city, or to be at the hub of things, which means that things happen and important decisions are made, or a hub airport is a large important one where people often change from one plane to another. In network analysis hubs refer to outgoing links. Kleinberg even proposes that hubs should link to good authorities.

In our case, Figure 2 shows that the department in the mid north represents three nodes of the highest hub values. Two nodes of the five nodes with the highest values are located in the department in the mid south of the graph.

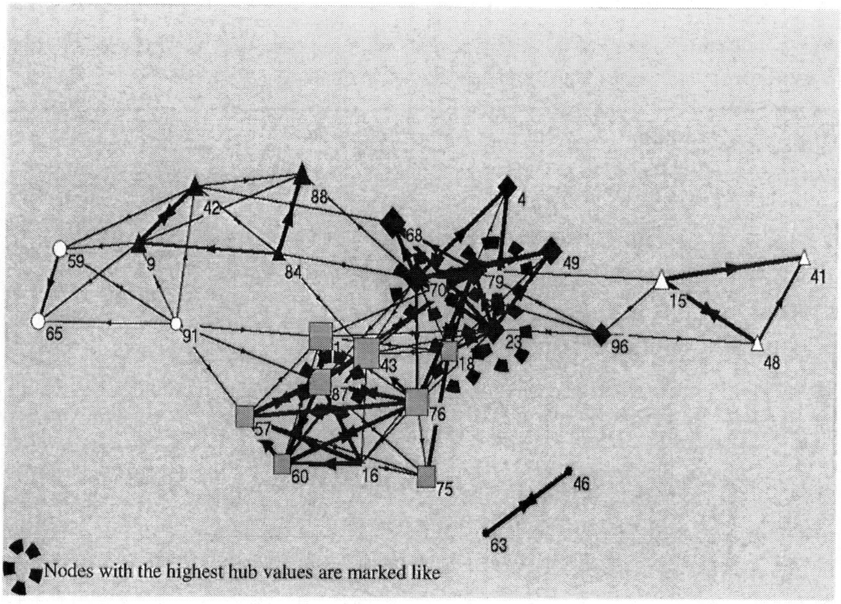

Figure 3. Communication network graph of the CoP (Q1), nodes with the highest hub values are marked.

Considering Table 2 we see that node 79 ranks first, with the value 0.4319, followed by node 70 (value 0.3855), and node 23 (value 0.3823). The node 18 ranks fourth with the value 0.3583 and the node 87 is on rank five with the value 0.2826. These identified nodes are important nodes concerning holding the connection, the linking to other nodes with influence of the best authorities, because each hub node is calculated based on the authority values of nodes of its outgoing links. E.g. the hub node 23 has to be calculated (because of its outgoing links) via

$$h(23) = a(1) + a(4) + a(18) + a(43) + a(49) + a(68) + a(70) + a(76) + a(79) + a(96),$$

Whereas the authority values come from the precedent authority iteration.

Table 2 also shows that the five nodes with the highest out-degree are also that ones with the highest hub values, but in reverse order. The argument for the different ranking for hubs and out-degree is analogous as for the authorities and in-degree up here.

Table 2.The hub ranks for the communication network (Q1) with the three steps of iteration to each node and their out-degree ranking.

Hubs	1. Iteration	2. Iteration	3. Iteration	Rank	Node	Out-degree
h (79)	0.4060	0.4259	**0.4319**	1	70	22
h (70)	0.3342	0.3784	**0.3855**	2	23	21
h (23)	0.3449	0.3746	**0.3823**	3	79	20
h (18)	0.3485	0.3555	**0.3583**	4	87	16
h (87)	0.2910	0.2853	**0.2826**	5	18	13
h (16)	0.2695	0.2749	**0.2738**	6	57	13
h (57)	0.2731	0.2694	**0.2673**	7	16	13
h (60)	0.2479	0.2427	**0.2395**	8	76	13
h (76)	0.2479	0.2426	**0.2372**	9	60	12
h (91)	0.2300	0.1599	**0.1420**	10	84	8
h (84)	0.1653	0.1276	**0.1190**	11	9	8
h (96)	0.1294	0.1045	**0.1033**	12	91	7
h (42)	0.1078	0.0758	**0.0648**	13	42	7
h (9)	0.1258	0.0496	**0.0323**	14	15	6
h (48)	0.0395	0.0140	**0.0121**	15	48	5
h (59)	0.0359	0.0121	**0.0061**	16	96	5
h (15)	0.0287	0.0049	**0.0028**	17	46	3
h (63)	0.0108	0.0002	**0.0000**	18	63	3
h (46)	0.0036	0.0001	**0.0000**	19	59	3

Next we consider the values of betweenness centrality for each node. Betweenness centrality is an indicator to measure the extent to which a particular person lies “between” other persons in the network. Nodes with high betweenness centrality are important for the shortest, i.e. fastest or most direct information flows but also vulnerable to having information flows disrupted by power plays or key individual leaving (Cross R. and Parker A. 2004). Figure 4 indicates such vulnerable points for our networks for Q1. The nodes 79, 87, 70, 91, and 84 take the first five ranks in order (see Table 3). Node 79 links many nodes within the department, connects the department in the east (besides 96), and the mid south (e.g. nodes 1, 18, 43, 96, etc). Many shortest paths to other nodes lead via node 79. This node is also the best hub node as its out-degree is rank 3, its authority is rank 7 and its in-degree is rank 6. Node 87 on rank 2 in the mid south of our network is an important broker in the sense of arranging the details between different parts or parties.

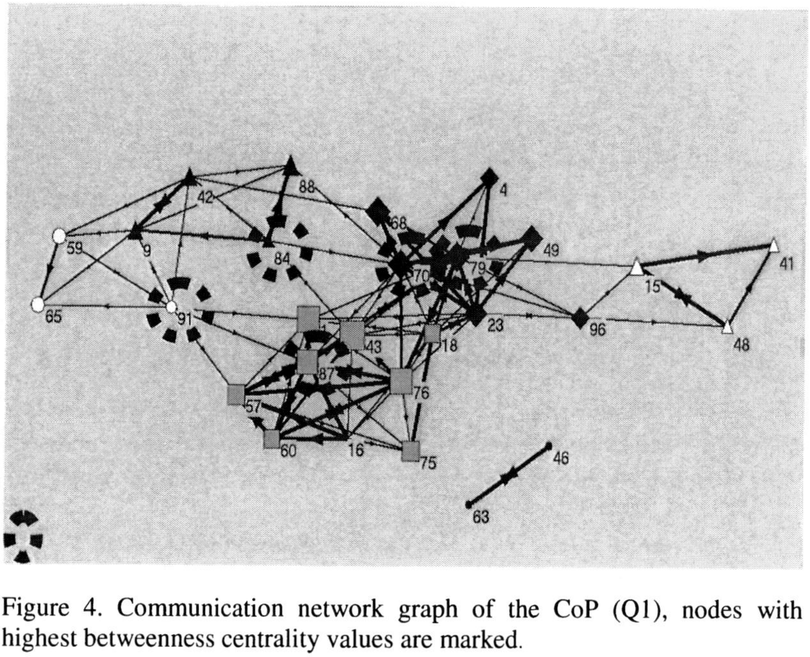

Figure 4. Communication network graph of the CoP (Q1), nodes with the highest betweenness centrality values are marked.

Nodes with the highest betweenness centrality values are marked like that.

Node 70 builds an important "bridge" inside the department and links to node 88 of the department in the North West. It follows node 79 as hub with the highest out-degree (but is on rank 10 as authority node). Important for keeping the network together are the nodes 91 and 84. Figure 4 shows that these nodes build bridges between several "continents" in our graph. Node 91 links two departments, the North West one and the mid south one, and is important for its colleagues within its department, the nodes 59 and 65. Node 84 connects its department with that one in mid south and therefore paths to this department lead over this node. Node 15 is on rank six, also very important, because for reaching node 41 the shortest path leads over node 15 and also to node 48, and the other way round.

Table 3. Betweenness centrality for communication network (Q1).

Node	Betweenness Centrality	Rank
79	0.10998	1
87	0.09990	2
70	0.07019	3
91	0.04723	4
84	0.03515	5
15	0.02422	6
96	0.02196	7
76	0.01679	8
18	0.01603	9
9	0.01228	10
42	0.00950	11
57	0.00924	12
23	0.00727	13
59	0.00421	14
48	0.00297	15
60	0.00028	16
16	0.00000	17
68	0.00000	18
75	0.00000	19
88	0.00000	20
1	0.00000	21
4	0.00000	22
41	0.00000	23
43	0.00000	24
46	0.00000	25
49	0.00000	26
63	0.00000	27
65	0.00000	28

The analysis presented here was also done for a so called **knowledge network.** As described in the methodology section, the members of the CoP were asked about their knowledge links: "**I understand this person's skills and knowledge independently of their relevance for my work**"(Q2). They had the options to mark the frequency: "none" (=0), "moderate" (=1), "good" (=2), "very good" (=3) which is visualised by means of the width of the arcs from one node to another. The frame of this article does not allow presenting the results here. But the results were relevant in out discussions with the responsible people for the CoPs (and were presented during the IAMOT conference). The analysis of this **knowledge network** allows tracing information flows.

Conclusions

The found results of our analysis were discussed with the people responsible for the CoPs in our considered organisation. Our results confirmed their assumptions regarding structure and dynamics of knowledge in the CoP. During our work, when we presented interim results, they helped us sharpening our approach, methods, and presentation. Information about specific tasks and skills, years in the job, duration of CoP membership, etc of a considered node are necessary for construing the detected roles of the nodes, because this quantitative analysis is only one specific view on the knowledge flow in the CoP and a will lead to distorted actions if overrated. The strategy, objectives of the CoP and the organisation, etc is not represented within this analysis. Therefore the quantitative results and the corresponding visualisations have to be positioned inside the context of quality aspects. The introduced methodological approach can be considered as valid in case it will be connected to a qualitative one.

Last but not least, our results also provided helpful input for the development of knowledge management measures in the CoP. A follow-up will be done after one year to see whether the knowledge flows have changed the functions of experts within the CoPs.

References

Boland, R. and Tenkasi, R. 1995. Perspective making and perspective taking in communities of knowing. In *Organization Science,* 6 (4), 350–372.

Brown, J.S. and Duguid, P. 1997. Organizational learning and communities-of-practice: toward a unified view of working, learning, and innovation. In *Organizational Science*, 2 (1), 40–57.

Brown, J.S. and Duguid, P. 1998. Organizing knowledge. In *California Management Review Spring*, 40 (3), 90–106.

Cross, R. and Parker, A. 2004. *The hidden power of social networks*. Harvard Business School Press, Boston.

Freeman, L. C. 1977. A set of measures of centrality based on betweenness. In *Socimetry* 40, 35-41.

Freeman, L. C. 1979. Centrality in social networks. 1. Conceptual clarification. In *Social Networks* 1, 215-239.

Freeman, L. C. 1980. The gatekeeper, pair-dependency and structural centrality. In *Quantity and Quality* 14, 585-592.

Kleinberg J., 1998. Authoritative Sources in a Hyperlinked Environment. In *Proc. 9th ACM-SIAM Symposium on Discrete Algorithms, Extended version in Journal of the ACM 46.*

Korfiatis, N. T.; Poulos, M.; and Bokos, G. 2006. Evaluating authoritative sources using social networks: an insight from Wikipedia. In *Online Information Review*, Vol 30, No. 3, pp 252 - 262.

Lave, J. and Wenger, E. 1991. *Situated learning - legitimate peripheral participation.* Cambridge University Press, Cambridge.

Lave, J. and Wenger, E. 1990. Situated Learning: Legitimate Peripheral Participation. Institute for Research on Learning, Palo Alto, CA.

McDermott, R. 1999. Why information technology inspired but cannot deliver knowledge management. In *California Management Review*, 41 103-17.

Wasserman, S. and Faust, K. 1994. *Social network analysis. Methods and applications*. Cambridge University Press, Cambridge.

Wenger, E. and Snyder, W. 2000. Communities of practice: the organizational frontier. In *Harvard Business Review*, January/February, 139–145.
Wenger, E.; McDermott, R.; and Snyder, W. 2002. *Cultivating Communities of Practice: A Guide to Managing Knowledge.* Harvard Business School Press, Boston, 2002
White, D. R. and Borgatti, St. P. 1994. Betweenness centrality measures for directed graphs. In *Social Networks* 16, 335-346.

DOES A FAVORABLE ENVIRONMENT HAVE A POSITIVE EFFECT ON UNIVERSITY TECHNOLOGY TRANSFER ACTIVITIES?: A CASE STUDY ON TWO TEXAS STATE UNIVERSITIES[14]

MICHI FUKUSHIMA

Faculty of Economics and Management, Tohoku University, 27-1 Kawauchi, Aoba-ku, Sendai 980-8576, Japan
michi@econ.tohoku.ac.jp

Universities are being increasingly recognized as important players in local economic development. In the U.S., a considerable number of universities have a mechanism for developing strong associations with the industry, such as Technology Licensing Offices (TLOs). Many researchers have pointed out that universities' technology transfer activities are associated with the regional business environment, particularly local companies' activities. Their discussion is implicitly based on the presumption that the business environment that is seemingly favorable to business or start-ups is always favorable to university technology transfer. We selected two state universities—the Texas A&M University (A&M) and the University of Texas at Austin (UT-Austin)—and compared their activities. Based on this comparative research, it became evident that a favorable business environment does not necessarily have a positive effect on the university's technology transfer activities; rather, the strategy that each university adopts is more important than the environmental factors.

Keywords: University, Technology transfer system, TLO, Environment, Regional development

[14] This research is supported by KAKENHI: Grant-in Aid for Scientific Research (B) 18330077

Introduction

Technology transfer from universities has been regarded as a knowledge source that leads to innovation. In the past years, some researchers have explained the differences in the performance of university technology transfer by focusing on organizational factors. On the other hand, other types of studies have been conducted that pay attention to business environment factors, such as Bahrami et al. (2000) and Saxenian (1994) who view Silicon Valley as an ecosystem. They presume that a business friendly environment contributes to technology transfer. In this paper, we question this presumption and subsequently provide another perspective.

Review and Research Methodology

There are some researchers that regard a region, such as the Silicon Valley, as an ecological system (Bahrami et al., 2000, Brown et al., 2000). The ecology consists of companies, local governments, universities, nonprofit organizations (NPO), etc., which are located in a region. These components of the ecology interact with each other and share the same customs and cultures, creating an environment for each other and also serving as a resource for each actor.

In particular, researches on Technology Licensing Offices (TLOs) suggest that the performance of university technology transfer should be determined by environmental factors (Siegel et al., 2003). These studies indicate that the environment surrounding universities affects the performance of university technology transfer. Moreover, they implicitly presume that a business-friendly environment is always favorable for the performance of university technology transfer.

In this paper, we will produce counterevidence to this presumption. In order to achieve this, we employ a comparative case study approach. More specifically, we will compare the development of the technology transfer system of two state universities in Texas, in the U.S.

Case Study

In this research, we compare the following two state universities: the Texas A&M University (A&M) and the University of Texas at Austin (UT-Austin). First of all, we need to ensure that both universities share some commonality and that they are comparable.

Feature of Each University

A&M and UT-Austin are both state universities in Texas, and they are both governed by the same s laws of the state of Texas. The two universities are approximately 100 miles apart. Neither of them has a medical center or a hospital, albeit A&M founded a Health Science Center in 1999 (Since the Center's scale is still small, it is negligible in our discussion). Based on these data, we can assume that these two universities are appropriate for a comparative study given that they share similar features in terms of scale and size (see Table1).

Table 1. Feature of universities

	A&M	UT-AUSTIN
Style	State University & Land grand university	State University
Departments (main department)	10 departments (agriculture, engineering, geo- science, animal)	15 departments (natural science, engineering, business school, pharmaceutical)
No. of faculties	About 2,500	2,653
No. of students	44,701	50,601
Graduate and others	8,049	12,007
Undergraduate	36,652	38,603

Business Environment Surrounding Both Universities

The environmental factors that surround the two universities are quite contrary to each other.

As indicated in Table 2, the size of the economy of Bryan-College Station MSA is about one-tenth the size of the economy in Austin-San Marcos MSA. Bryan-College Station's business environment is uncultivated in comparison to the Austin area. From the perspective of start-ups, Austin has a far superior business environment in addition to appearing more favorable for technology transfer.

Table 2. Comparative environments between College Station and Austin (2002)

	College Station	Austin
Population	152,415 (30% are students)	**1256,000**
Number of companies (Number of jobs)	3,266 (78000) *mainly government related	**31,402 (671700) *mainly high-tech, such as IBM, AMD, Applied Materials Inc., Dell Computers, and Motorola**
Number of high-tech companies	84	**About 200**
Number of venture capital firms	Few	**About 60**
Number of accountants	196	**969**
Number of attorneys	256	**6,828**

History of the Technology Transfer System of Each University

In retrospect, their histories of technology transfer suggest that A&M and UT-Austin have experienced the same problems but have followed different solutions.

Origin and mission of each university: Era of laissez-faire

A&M was originally founded with the objective of contributing to regional economic development through science and technology knowledge. A&M was opened on October 4, 1876, as the

Agricultural and Mechanical College of Texas. The school owes its origin to the Morrill Act of 1862. A&M includes state agencies, such as Texas Agricultural Experiment Station (TAES) and Texas Engineering Experiment Station (TEES) that conduct research for public and practical purposes. Most of the faculty members of the engineering department of A&M hold posts as researchers of TEES. However, these and other state agencies within A&M were not necessarily active before the 1980s. Even people who were working for state agencies were not aware of their existence.

Since its foundation in 1883, UT-Austin's most important mission has been education and research, and it has aspired to become one of the top ten national research universities. However, UT-Austin has also informally produced some spin-off companies, such as Tracor, Inc. (founded in 1955) and National Instruments (founded in 1976). Tracor grew into a Fortune 500 company and gave birth to many spin-offs (Brett et al., 1991). These have no doubt contributed to Austin's economic development. They did not receive any support from the university when they began their business activities. It is fair to state that although everything happened voluntarily without any assistance from the university, the university was tolerant of their activities

Policies in the 1980s: The time of state-driven technology transfer

In the 1980s, people in Texas began to realize that the Texas economy, which had heavily depended on the oil industry, faced a transition and needed to be diversified. Further, they began to regard the university as an important source of knowledge for the creation of new industries. Also in the 1980s, certain acts that facilitated technology transfer from universities in Texas were enacted by the State.

In 1979, William P. Clements Jr. was elected Governor of Texas. During his campaigns, Clements vowed to make the state government more business-like. In April 1981, the Texas 2000 Commission was established under his leadership, and research was conducted outlining nine areas that would be important to the Texas economy in the future. One of the nine areas—science and technology—was chosen as one of the most important areas that needed to be addressed on a timely basis.

The subsequent Governor, Mark White, was equally eager to facilitate R&D activities in Texas. Following the Texas 2000 Commission's recommendations, White set up the Science and Technology Council in the Office of the Governor by an Executive Order in June 1984 to discuss economic development and obtain opinions from a variety of people. In 1985, the House Committee on Science and Technology was set up in the Legislature. Through a series of policies, the Science and Technology Council began to focus on the role of universities. This Council drafted certain recommendations on university technology transfer.

The initial efforts at a full-fledged technology transfer began in the early 1980s at A&M. Although they had had some agencies, such as TEES and TAES, they were not necessarily active at the time[15]. However, this began to change when W. Arthur (Skip) Porter took over as director of TEES in 1980. He stressed on the importance of spurring entrepreneurship and the manner in which to convince the university research and development was needed[16]. On receiving Dr. Porter's proposal, the commission proposed to "establish a state Business Development Corporation, which would fund and offer professional assistance for the development or innovation from universities, state agencies, individuals and small companies[17]". This proposal was included in the final version of the Texas 2000 project recommendations in 1982.

Based on these recommendations, A&M established a new program named Institute for Venture for New Technology (INVENT) as part of TEES under the control of the board of regents of the Texas A&M University System. The purpose of INVENT was "to provide technical assistance in the evaluation

[15] Armstrong (1984)
[16] Armstrong (1984)
[17] Texas Past and Future: A Survey, p.165

and development of new products and processes for small business enterprises and individual entrepreneurs.

While INVENT did a considerable amount of work in all regards, it was unable to achieve much due to strong resistance from the faculty members[18]. When Dr. Porter resigned from the position of the Director of TEES in 1985, the program was also reorganized and merged into the Technology Business Development Division (TBD) in 1987 through legislative authorization. TBD was formed as a new entity for administering INVENT.

One of the important missions of TBD was to commercialize research results in the university. In fact, TBD indicated that it would assist start-ups in a variety of ways, including feasibility studies, market research, business plan development, product development assistance, intellectual property information, and information on financing alternatives. TBD not only supported the management of existing small businesses but also dealt with contracts with the Federal Government. TBD secured contracts from parties such as the U.S. Economic Development Administration (EDAM) and the Texas Higher Education Coordinating Board[19].

Despite its intentions, TBD was unable to achieve sufficient success to secure continuous funding. The State of Texas withdrew funding in 1992 after the original 5-year window of funding had closed. Instead, A&M's TLO was founded in the same year; the year Dr. Terry Young, who had worked for TBD before, became the director of the organization. The TLO lobbied to revive TBD as part of its activities[20], and as a result of their efforts, the Texas State Legislature passed the bill in 2001 that gave A&M the right to create a Center for Technology Development and Transfer under

---

[18] Interview with Dr. Jane Armstrong (former director of INVENT, Texas A&M university) September 10, 2002

[19] Interview with Dr. Terry Young (Former director of the TLO of the Texas A&M University) July 17, 2002

[20] Interview with Dr. Terry Young (Former director of the TLO of the Texas A&M University) July 17, 2002

Chapter 53 of the Education Code. If the A&M regents approved such a center, A&M would have an entity for new technology transfer, which would incorporate TLO activities.

These efforts at A&M, particularly INVENT, had an effect on UT-Austin. Impressed by INVENT, Dr. Earnest F. Gloyna, Dean of the Engineering Department of the university, drafted a proposal to Governor White requesting a center for technology transfer at UT-Austin. Initiated by Dr. Gloyna's proposal and lobbying, the Center for Technology Development and Transfer (CTDT) at UT-Austin was established by the 69th Texas legislature in 1985.

In 1986, UT-Austin created the CTDT, which facilitated commercialization of university research by university entrepreneurs. While the organization was actually a political state entity, it received its budget from the College of Engineering primarily because the Dean of the College was an advocate in the founding of the organization.

The CTDT had learned from the experiences of INVENT at A&M. INVENT's target customers were existing small companies, for which it provided support for technology transfer and development. However, working with existing small businesses required excessive time and a considerable amount of labor. Learning from the experiences of INVENT, the CTDT focused on commercializing university research and helped the researchers in the university who wished to commercialize their inventions. The CTDT provided a wide variety of services to these researchers, largely because it was a new organization and did not have a model to follow.

The first task of the CTDT was to educate and convince campus organizations such as faculty committees, the patent committee, and the Dean's Council to use the CTDT. They informed these entities about the CTDT and its functions. Subsequently, they guided faculty who wished to create a new business or invention by advising them on intellectual property and conducting marketing research. Sometimes they created associations between the faculties of different departments who required each other's technology.

The CTDT helped faculty members at UT-Austin to commercialize their technology. Desk Top Manufacturing Corporation (DTM) was one of the success stories of the CTDT. DTM was formed in 1987 by Harold Blair and Dr. Paul F. McClure, who licensed the patent from the University of Texas System. The CTDT played an indispensable role as the intermediary in making DTM a success, introducing Harold Blair to venture capitalists. In this supporting process, the CTDT, with an aim to maximize profits, sometimes was in conflict with UT-Austin, which, in general, was hesitant to take risks[21].DTM grew rapidly and went public in 1997. It was the second largest company in the 3-D industry at the time when they introduced their initial public offering, following 3-D Systems Corp. of California. In 2001, DTM merged with 3-D Systems Corp., becoming the largest firm in the industry.

Despite some achievements, budget problems brought the CTDT to a deadlock in 1989. Although the CTDT was an organization created by the Legislature and its activities were appreciated, the organization was too innovative for the university to be able to keep up with it. At times, the CTDT had conflicts with the Vice President of UT-Austin. UT-Austin was unable to accept the CTDT's risk-taking activities[22].

As a result of these serious conflicts, the CTDT lost the support of the university. University officials turned down the CTDT's request to be in the school's budget. Consequently, the CTDT was forced to secure money from other sources. In May 1989, legislators approved an annual $250,000 for the CTDT for the next two years. However, this money was obtained from grants that were given exclusively to energy-related projects. Therefore, this

---

[21] Interview with Meg Wilson (Lecture on IC2 at the University of Texas in Austin) February 14, 2002

[22] Interview with Meg Wilson (Lecture on IC2 at the University of Texas in Austin) February14, 2002

kind of money restricted the scope of the CTDT's activities[23]. While the CTDT still existed on paper, the organization was unable to garner funding from sources other than the state oil overcharge money.

Technology transfer through patents in the 1990s

In the 1980s, the State of Texas began recommending each university to set up TLOs. In addition, under Governor White's leadership, in 1984, the Science and Technology Council began focusing on the role of universities and drafted some recommendations on university technology transfer. First, the university should positively engage in technology transfer and have a support organization for it. This recommendation was realized in the form of INVENT, TBD, and the CTDT. Second, every university should have a clear and simple intellectual property policy to help accelerate technology transfer from the university. Finally, universities should be permitted to hold equity ownership as an incentive. These recommendations were supported by not only politicians but also industry associations such as AeA and Technology Industry Task Force.

However, the setting up of technology transfer offices, in reality, only took place 11 years after the Bayh-Dole Act was passed in 1980, and eight years later, the A&M system founded their TLO. Dr. Terry Young, former member of A&M's TBD, was chosen as the executive director of the TLO. The TLO's mission was "to protect the proprietary research results of the Texas A&M University System through patents and copyrights, and to transfer those intellectual properties to industry by royalty-bearing license agreements for commercial products, economic development, and public benefit[24]."Under the leadership of Dr. Young, the TLO has performed well in comparison with other universities in the U.S. According to research by The Chronicle of Higher Education in 2002, A&M ranks eighth in terms of licensing income per dollar of

[23] Austin American Statesman, June 1, 1989
[24] TLO at A&M home page (http://tlo.tamu.edu/content.cfm?Section=INFO)

research spending[25]. TEEE and TAES have played central roles in technology transfer, contributing to 85% of the total disclosures at A&M.

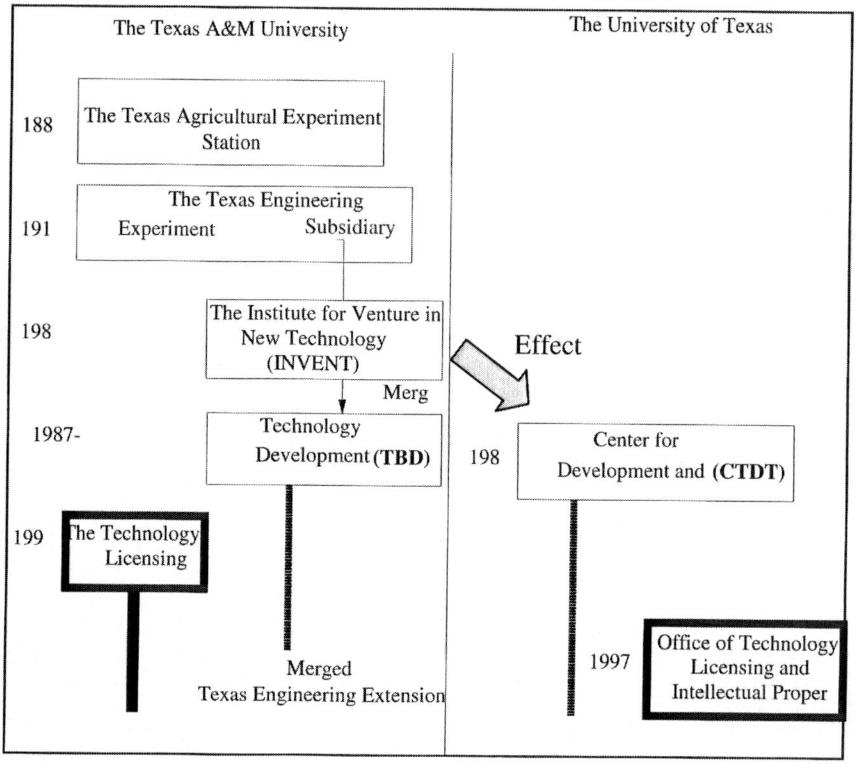

Figure 1. Main organizations that engage in technology transfer in each university

With regard to patents, for a long time, UT-Austin had an atmosphere that discouraged researchers from filing for patents. People who were against technology transfer insisted that pursuing patents was not a part of UT-Austin's academic mission. In addition, owing to the fact UT-Austin is a tax-supported state university; many believed that selling intellectual property generated from research on campus was an improper activity for its employees. UT-Austin has an Intellectual Property Committee,

[25] The Chronicle of Higher Education, July 19, 2002, A24

which meets once a month to consider recommendations to pursue patent filings and research results. However, it was doubtful how well the committee actually worked. Additionally, the $20,000 cost of filing per patent was considerably expensive, and hence, the university was generally opposed to patent filings[26].

UT did not have a full-time employee working on the licensing program until 1991. Only one employee, Patricia Ohlendorf, UT-Austin's vice provost and UT General Counsel, handled licensing on a part-time basis. Intellectual property-related jobs were regarded as one of the functions of the provost's office. In 1991, UT-Austin hired its first full time licensing specialist; by 1995, there were three employees working with the program[27]. The culture that opposed technology transfer began to change gradually in the 1990s, becoming evident around 1997. Finally, UT established the Office of Technology Licensing & Intellectual Property (OTL) in September 1997, as a formal department of the university.

The first driving forces for the change were the new faculty members who came to UT. Creating the Biomedical Engineering Department and bringing in over 20 new faculty members brought a significant cultural change to the school. In addition, UT-Austin organized the Center for Nano & Molecular Science and Technology in 2000. The center hired 23 new faculty members who came to UT-Austin expecting the services for technology transfer from the university[28]. The second driving force was Austin's business communities such as business people, the alumni of UT, and the Chamber of Commerce. They expected UT to encourage and facilitate technology licensing and transfer as the economic situation of Austin improved.

---

[26] Austin Business Journal, November 10, 1997

[27] Austin Business Journal, November 10, 1997

[28] Interview with Renne Mallett, (Former director of the OTL, the University of Texas in Austin) September 6, 2002

Organizational Structure of TLOs

As of 2002, both universities have technology licensing organizations. However, the position of these organizations and their manner of functioning were considerably different. A&M has a centralized structure while UT-Austin has a decentralized structure. Currently, A&M's TLO plays a central role in technology transfer. The TLO at A&M reports to the Vice Chancellor of Business Services, which deals with all the technology created in the A&M System. In 2002, 19 people worked at the A&M's TLO. On the other hand, the UT System has decentralized TLOs. The System has 14 components, six of which are medical health components, and each component has its own TLO. Therefore, the OTL at UT-Austin is positioned at a lower level than the TLO at A&M. and only covers technologies in UT-Austin. Just seven employees are working for this organization in 2002.

Further, in terms of budgeting also, both universities have different systems. The TLO at A&M shares the license fees with inventors and the university at a ratio of 42.5:42.5:15, respectively. In other words, the TLO at A&M is a self-supporting organization[29]. On the other hand, the OTL at UT-Austin is a so-called self-supported organization. License revenues from patentable subject matters are shared equally between the inventive team and the university. The OTL at UT-Austin receives reimbursement for their operation costs from the university budget. The university determines the budget of the OTL annually, based on the previous year's expenditure and the number of employees. In 2002, they received $1.1 million. This amount will increase in accordance with any increase of personnel[30]. At UT-Austin, the revenue of the OTL does not reflect their production directly. With regard to budgeting, A&M has a system to motivate the TLO in that the more they produce, the more they will receive in budgeted funds.

---

[29] Interview with Dr. Terry Young, (Former director of the TLO of the Texas A&M University), July 17, 2002

[30] E-mail interview with Renee Mallett, (Former director of the OTL of the University of Texas in Austin)

Performance of Technology Transfer

In this section, we address the performance of both the organizations. In terms of disclosure, A&M is more active than UT-Austin. With respect to the gross royalty, licensing, and options generating royalties for the fiscal year, A&M has annually increased its revenue (Table 3).

Table3. Licenses and options generating royalties for a fiscal year

YEAR	A&M	UT-AUSTIN
1991	-	70
1992	-	90
1993	43	-
1994	57	-
1995	79	-
1996	99	21
1997	112	24
1998	137	26
1999	155	24
2000	203	21

Although UT-Austin has also shown a gradual increase, the difference between A&M and UT-Austin has amplified over the years. With regard to the total IP income including patents, copyrights, and licensing agreements, A&M shows a steady increase every year. UT-Austin's level increased in 1996, but has remained constant since 1997 (Figure 2).

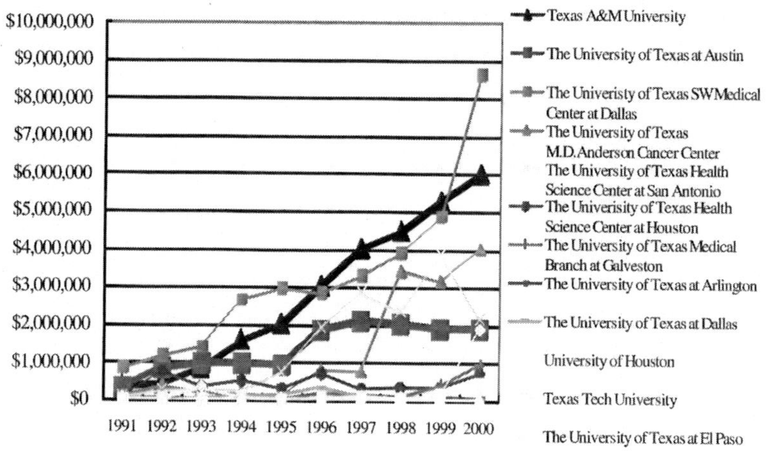

Source: the Texas Higher Education Coordinating Board (http://www.thecb.state.tx.us/)

*Intellectual property include patents, copyright, and licensing agreement. Total income derived from royalities, licensing or other transactions related to intellectual property derived from research.

Figure 2. IP income of Texas Universities

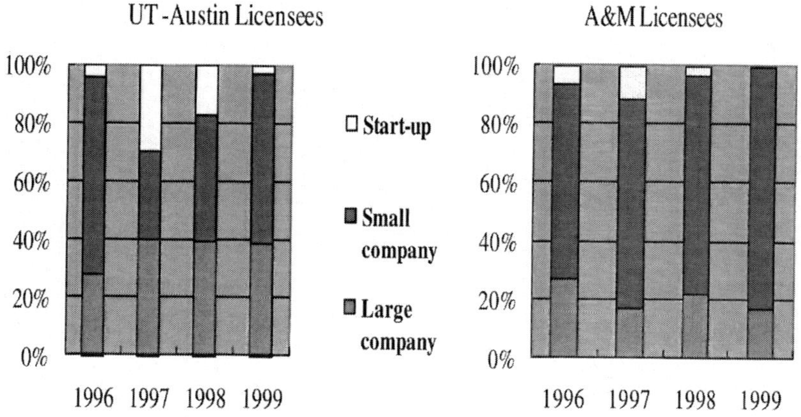

Figure 3. Licensees of each university

The universities exhibited differences in licensees 1996-1999. The licensees of UT-Austin are small, large, and start-up firms. A&M, on the other hand, works with small companies (Figure 3). With respect to the licensees' distribution by region, A&M has at least one licensee in 43 states of the U.S.[31], while UT-Austin has licensees in only 17 states[32]. In addition, A&M has licensees in 77 countries, while UT-Austin has licensees in only 7 countries. In interpreting these data, we must consider the differences in the history as well as the TLO performance of both universities.

Analysis

Considering the history of both universities, one can observe that both universities have had different methods of technology transfer. This leads to the following questions: What is the difference between the two universities? Further, what caused this difference between the two universities? It is evident that both systems developed in accordance with their own environment and organizations.

The Driving Force Behind the Changes at Each University

The historical facts suggest differences in each university's attitude toward technology transfer. Both the universities have had conflicts between academia and business. These are unavoidable when universities attempt to engage in technology transfer. However, what made A&M different from UT-Austin was leadership. A&M always possessed an agent of change or visionaries, such as Dr. Porter or Dr. Young. They provided novel ideas, pitched for funding from the state, and handled the intra-organizational conflicts. Their activities should be appreciated and acknowledged, even if they were not necessarily successful.

On the other hand, UT-Austin lacked innovation and strong leadership. With regard to their technology transfer policy, they merely followed A&M's policies or made alterations under external pressure. Moreover, while UT-Austin could have taken advantage of many of the opportunities presented, it lacked leadership.

---

[31] The Texas A&M TLO, licensing data

[32] The University of Texas at Austin, the Office of Technology Licensing and Intellectual Property, HP. (http://www.utexas.academic/otl/)

Organizational Character of Both TLOs

A&M's TLO is a system-level organization that deals with the technology from the entire A&M system. It has 19 employees who work with a variety of companies or institutes inside and outside of Texas as well as the U.S. Since A&M's TLO is a self-sufficient organization, it is considerably motivated to perform. On the other hand, the OTL of UT-Austin has seven employees that only deal with the technology from UT-Austin. Although the organization has been working hard, it still lacks power and its growth is under the university's control.

The Technology Transfer Strategy Adopted by Each University

A&M adopted the strategy of centralizing all its resources and power into the TLO in order to compensate their disadvantages. Focusing on patents was a good strategy for A&M to overcome the disadvantage of distance. In addition, A&M carried this out in a systematic manner. The university installed TLO at a higher position. Hence, the TLO served as a pump, pushing out their technology. The activities of A&M's TLO extended over a large geographical area, not only inside the U.S. but also abroad.

On the other hand, the UT System has a decentralized structure for technology transfer. Each campus has its own TLO, and UT-Austin is just one of them, which lacks a medical school or hospital. UT-Austin has not given much power and autonomy to the OTL. Therefore, the OTL's activities are limited by UT-Austin's authority through budget and personnel. In this sense, the OTL did not serve as a pump; rather, it was more like a waterway through which technology was merely transferred.

It is worth considering whether UT-Austin has conceived a consistent and clear policy for technology transfer. Further, UT-Austin has not displayed any strong and successful leadership with regard to technology transfer. In retrospect, the technology transfer attempts implemented in UT-Austin did not originate inside the campus; instead, they were results of external pressure.

What Made Them Different?

Historically, A&M was founded as a land grant university with its supreme mission to contribute to regional development. Due to meager resource and no player in region, A&M had no way other than being center of regional development and concentrating resource. As a result, A&M has grown into an industry-academic complex.

On the other hand, Austin has been the capital city as well as a university city for a long time, with few big businesses. Austin succeeded in recruiting the Microelectronics and Computer Technology Corporation (MCC) in 1983[33]. In 1987, Austin was also chosen as the place for the headquarters of Semiconductor Manufacturing Technology (SEMATECH). In addition, some other projects of large companies in the 1980s, such as the IBM AIX project (1984) and the TI ASC project (1966), attracted many engineers and researchers to Austin[34]. Accompanying them were service industries and others that supported business creation, which began to accumulate in Austin. As a result, a favorable infrastructure for business was formed.

Even when provided a favorable environment for business creation with many business opportunities, UT-Austin did not actively prepare or organize their resources for business. One of the reasons for this is that UT-Austin's historical mission had not necessarily been regional development. Their main mission has always been education and to rank among the top ten universities in the U.S. Second, high-tech companies in Austin belonged mainly to the software, semiconductor, and PC/peripheral industries. These industries face fast-pace technological change and need agility to capitalize on market opportunities. From the perspective of business, technology transfer from the universities takes excessive time; thus, the businesses were hesitant in utilizing the resources in universities. Third, there should have been more flexible and

---

[33] Gibson et al. (1994)

[34] Fukushima (2004a,b)

hidden linkages between the faculty members of the university and local businesses. Before the 1980s, some faculty members created new companies, such as Tracor and National Instruments, while they did not receive any help from UT-Austin. They worked for their businesses informally and finding out business supporters by themselves. These kind of informal networks have inhibited the construction of formal linkage between university and business. While it would be difficult to verify this, it may be possible to collect some anecdotes.

Concluding Remarks

Based on this in-depth case study, we can state that a favorable business environment does not necessarily facilitate technology transfer. On the contrary, internal factors of a university, such as strategy, culture, and organization, are more important than environmental factors. In order to generalize this finding, additional research needs to be conducted. Our research can be regarded as the first step toward understanding university technology transfer.

References

Armstrong, J (1984) *Oral History Collection at Texas A&M University.*

Audretsch, D.B. and M.P. Feldman (1996) "Spillovers and the geography of innovation and production". *The American Economic Review*, 86(3), 630–640.

Bahrami, H. and S. Evans (2000) "Flexible recycling and high-technology entrepreneurship". In *Understanding Silicon Valley: The Anatomy of an Entrepreneurial Region*, M Kenney (ed.), 165–189. Stanford: Stanford University Press.

Brett.A.M.,Gibson,D.V. and Smilor, R.W. (1991) *University Spin-off Companies: Economic Development, Faculty Entrepreneurs and Technology Transfer*, Savage MD: Rowman and Littlefield Publishers.

Brown, J.S. and P. Duguid (2000) "Mysteries of the region: Knowledge dynamics in Silicon Valley". In *The Silicon Valley Edge: A Habitat for Innovation and Entrepreneurship*, Miller et al. (eds.), 16–39, Stanford: Stanford Business Books.

Henderson, R., A.B. Jaffe and M. Trajtenberg (1998) Universities as a source of commercial technology: A detailed analysis of university patenting, 1965-1988, *The Review of Economics and Statistics*, 80(1), 119–127, Cambridge: MIT Press.

Fukushima, M (2004a) "Groundbreaking years of establishing the technology transfer system at the University of Texas at Austin." *Annual Report of the Economics Society*, Tohoku University, 66(1), 1–20.

Fukushima, M (2004b) "The development of the technology transfer system at the University of Texas at Austin."*Annual Report of the Economics Society*, Tohoku University, 66(2), 315-334.

Gibson, V.D, and E.M. Rogers (1994) *R&D Collaboration on Trial.* Cambridge: Harvard Business School Press.

Jaffe, A (1989) "Real effects of academic research", *American Economic Review*, 79(5), 957–970.

Jaffe, A., M.Trajtenberg and R Henderson (1993) "Geographic localization of knowledge spillovers as evidenced by patent citation." *Quarterly Journal of Economics,* 108(3), 577–598.

Saxenian, A. (1994) *Regional Advantage.* Cambridge: Harvard University Press.

Siegel, D.S., D. Waldman and A. Link (2003) "Assessing the impact of organizational practices on the relative productivity of university technology transfer offices: An exploratory study", *Research Policy*, 32(1), 27–48.

Thursby, G.T, R. Jensen and M.C. Thursby (2001) "Objectives, characteristics and outcomes of university licensing: A survey of major U.S. universities". *Journal of Technology Transfer*, 26(2), 59–72.

Trajtenberg, M., R.Henderson and A. Jaffe (1997) "University versus corporate patents: A window on the basicness of innovation". *Economics of Innovation and New Technology*, 5(1), 19–50.

Wilson, M, and S Szygenda. (1991) "Promoting university spin-offs through equity participation". In *University Spin-off Companies*, Brett, A., G. Gibson and R. Smilor (eds.). Maryland: Rowman and Littlerfield Publishers, Inc.

Zucker LG, MR Darby and MB Brewer (1998). "Intellectual human capital and the birth of U.S. biotechnology enterprises". *American Economic Review*, 88(1), 290–306.